Robot Programming: A Guide to Controlling Autonomous Robots

Cameron Hughes
Tracey Hughes

que

800 East 96th Street
Indianapolis, Indiana 46240

ROBOT PROGRAMMING: A GUIDE TO CONTROLLING AUTONOMOUS ROBOTS

ISBN-13: 978-0-7897-5500-1

ISBN-10: 0-7897-5500-9

Library of Congress Control Number: 2015955656

2 16

Trademarks

Warning and Disclaimer

Special Sales

For information about buying this title in bulk quantities, or for special sales opportunities (which may include electronic versions; custom cover designs; and content particular to your business, training goals, marketing focus, or branding interests), please contact our corporate sales department at corpsales@pearsoned.com or (800) 382-3419.

For government sales inquiries, please contact governmentsales@pearsoned.com.

For questions about sales outside the U.S., please contact intlcs@pearson.com.

Editor-in-Chief
Greg Wiegand

Executive Editor
Rick Kughen

Senior Acquisitions Editor
Laura Norman

Development Editor
William Abner

Technical Editor
John Baichtal

Managing Editor
Sandra Schroeder

Project Editor
Mandie Frank

Copy Editor
Geneil Breeze

Indexer
Ken Johnson

Proofreader
Gill Editorial Services

Editorial Assistant
Cindy Teeters

Cover Designer
Chuti Prasertsith

Compositor
Bronkella Publishing

CONTENTS AT A GLANCE

CONTENTS

ABOUT THE AUTHORS

Cameron Hughes is a computer and robot programmer. He holds a post as a Software Epistemologist at Ctest Laboratories where he is currently working on A.I.M. (Alternative Intelligence for Machines) and A.I.R. (Alternative Intelligence for Robots) technologies. Cameron is the lead AI Engineer for the Knowledge Group at Advanced Software Construction Inc., a builder of intelligent robot controllers and software-based knowledge components. He holds a staff appointment as a Programmer/Analyst at Youngstown State University.

Tracey Hughes is a senior software and graphics programmer at Ctest Laboratories and Advanced Software Construction Inc. where she develops user interfaces and information and epistemic visualization software systems. Her work includes methods of graphically showing what robots and computers are thinking. She is on the design and implementation teams for the East-Sidaz robots at Ctest as well.

Both Cameron and Tracey Hughes are members of the advisory board for the NREF (National Robotics Education Foundation) and members of the Oak Hill Collaborative Robotics Maker Space. They are project leaders of the technical team for the NEOACM CSI/CLUE Robotics Challenge and regularly organize and direct robot programming workshops for the Arduino, Mindstorms EV3, LEGO NXT, and RS Media robot platforms. Cameron and Tracey are two of the authors of *Build Your Own Teams of Robots with LEGO® Mindstorms® NXT and Bluetooth*, published by McGraw-Hill/TAB Electronics, January 2013. They have written many books and blogs on Software Development and Artificial Intelligence. They've also written books on multicore, multithreaded programming, Linux rapid application development, object-oriented programming, and parallel programming in C++.

Dedication

We dedicate this book to all those open source robot maker spaces that in spite of humble and meager resources continue to toil against the improbable and do amazing things with robots.

ACKNOWLEDGMENTS

We are greatly indebted to Valerie Cannon who played the role of "on location" robo-journalist and photographer for us at the 2015 DARPA Robotics Search and Rescue Challenge at the Fairplex in Pomona, California.

We would like to thank our two interviewees for our "Bron's Believe It or Not" interviews. We also thank Ken Burns from Tiny Circuits of Akron, Ohio, who provided us with a personal tour of his Arduino manufacturing space and endured our probing interview questions. Portions of the material on Arduino robotics hardware, especially the Phantom X Pincher Robot Arm, would not have been possible without the time and interview given to us from Kyle Granat at Trossen Robotics.

We are also indebted to the NEOACM CSI-Clue robotics challenge team who acted as a sounding board and early test bed for many of the robot example programs in this book. We are fortunate to be part of Ctest Laboratories, which provided us with unfettered access to their East Sidaz and Section 9 robots. The East Sidaz and Section 9 met every challenge we could throw at them. A special thanks to Pat Kerrigan, Cody Schultz, Ken McPherson, and all the folks at the Oak Hill Collaborative Robotics Maker Space who allowed us to subject them to some of our early robot designs. A special thanks to Howard Walker from Oak Hill Collaborative who introduced us to the Pixy camera. Thanks to Jennifer Estrada from Youngstown State University for her help with the Arduino-to-Bluetooth-to-Vernier magnetic field sensor connection and code. A special thanks goes to Bob Paddock for offering his insight and expertise on sensors and giving us a clear understanding of the Arduino microcontroller. A shout-out to Walter Pechenuk from IEEE Akron, Ohio, chapter for his subtle, cool, and calm interaction and responses as we went on endlessly about our approach to autonomous robotics. Further, this simply could not have been written without the inspiration, tolerance, and indirect contribution of many of our colleagues.

WE WANT TO HEAR FROM YOU!

As the reader of this book, *you* are our most important critic and commentator. We value your opinion and want to know what we're doing right, what we could do better, what areas you'd like to see us publish in, and any other words of wisdom you're willing to pass our way.

We welcome your comments. You can email or write to let us know what you did or didn't like about this book—as well as what we can do to make our books better.

Please note that we cannot help you with technical problems related to the topic of this book.

When you write, please be sure to include this book's title and author as well as your name and email address. We will carefully review your comments and share them with the author and editors who worked on the book.

Email: feedback@quepublishing.com

Mail: Que Publishing
           ATTN: Reader Feedback
           800 East 96th Street
           Indianapolis, IN 46240 USA

READER SERVICES

Register your copy of *Robot Programming* at quepublishing.com for convenient access to downloads, updates, and corrections as they become available. To start the registration process, go to quepublishing.com/register and log in or create an account*. Enter the product ISBN, 9780789755001, and click Submit. Once the process is complete, you will find any available bonus content under Registered Products.

*Be sure to check the box that you would like to hear from us in order to receive exclusive discounts on future editions of this product.

INTRODUCTION

ROBOT BOOT CAMP

 caution

We who program robots have a special responsibility to make sure that the programming is safe for the public and safe for the robots. The safety of robot interaction with humans, animals, robots, or property is a primary consideration whenever a robot is being programmed. This is true for all kinds of robot programming and especially true for programming autonomous robots, which is the kind of robot programming that we explain in this book. The robot commands, instructions, programs, and software presented in this book are meant for exposition purposes only and as such are *not* suitable for safe public interaction with people, animals, robots, or property.

A serious treatment of robot safety is beyond the scope of this introductory book. Although the robot examples and applications presented in this book were tested to ensure correctness and appropriateness, we make no warranties that the commands, instructions, programs, and software are free of defects or error, are consistent with any particular standard of merchantability, or will meet your requirements for any particular application.

The robot code snippets, programs, and examples are meant for exposition purposes only and should not be relied on in any situation where their use could result in injury to a person, or loss of property, time, or ideas. The authors and publisher disclaim all liability for direct or consequential damages resulting from your use of the robots, commands, instructions, robot programs, and examples presented in this book or contained on the supporting website for this book.

Robot Programming Boot Camp

Welcome to *Robot Programming: A Guide to Controlling Autonomous Robots*. This robot programming "boot camp" ensures that you have all the information needed to get started. We have built and programmed many types of robots ranging from simple single-purpose robots to advanced multifunction autonomous robot teams and have found this short robot programming boot camp indispensable for those who are new to programming robots or who want to learn new techniques to program robots.

Ready, Set, Go! No Wires or Strings Attached

There are two basic categories for robot control and robot operation as shown in Figure I.1.

The telerobot group represents robot operations that are remotely controlled by a human operator using some kind of remote control device or puppet mode. Some remote controls require a tether (a wire of some sort) to be physically connected to the robot, and other types of remote control are wireless (for example, radio control or infrared).

The autonomous robot group represents the kind of robot that does not require a human operator. Instead, the robot accesses a set of instructions and carries them out autonomously without intervention or interruption from a remote control.

In this book, we focus on the autonomous group of robot operations and robot programming. Although we often discuss, explain, and contrast telerobots and autonomous robots, our primary focus is on introducing you to the basic concepts of programming a robot to operate and execute assigned tasks autonomously.

As you see in Chapter 9, "Robot SPACES," there are hybrids of the two types of robot control/operation with different mixes and matches for operation strategies. You are introduced to techniques that allow for mixing and matching different robot control strategies.

caution

Although *Robot Programming: A Guide to Controlling Autonomous Robots* does not assume that you have any previous experience programming robots, to get the most out the book it is assumed that you are familiar with basic programming techniques in standard programming languages such as Java or C++. While the book does present all the final robot programs in Java or C++, the basic robot instruction techniques and concepts are presented with diagrams or plain English first. The book also introduces you to approaches to program design, planning, and analysis such as RSVP (Robot Scenario Visual Planning) and REQUIRE (Robot Effectiveness Quotient Used in Real Environments) .

note

All robot instructions, commands, and programs in this book have been tested on ARM7, ARM9 microcontroller-based robots as well as on the widely available and popular LEGO NXT and EV3-based robots. All other robot-based software used in this book was tested and executed in Mac OSX and Linux environments.

Figure I.1
The two basic categories of robot operation

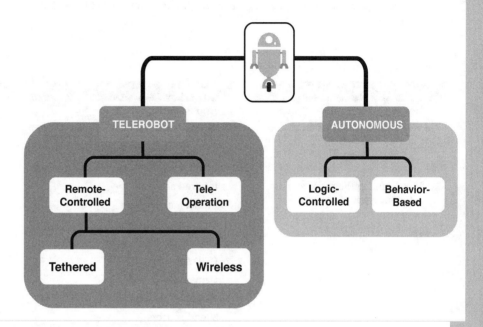

Boot Camp Fundamentals

Five basic questions must be answered prior to any attempt to program a robot:

1. What type of robot is being considered?

2. What is the robot going to do?

3. Where is the robot going to do it?

4. How is the robot going to do it?

5. How will the robot be programmed?

Many beginner and would-be robot programmers fail to answer these basic questions and end up with less than successful robot projects. Having the answers to these fundamental questions is the first step in the process of getting any kind of robot to execute the assigned task. In *Robot Programming: A Guide to Controlling Autonomous Robots* we demonstrate how these questions and their answers are used to organize a step-by-step approach to successfully instructing a robot to autonomously carry out a set of tasks.

Core Robot Programming Skills Introduced in This Book

In this book, we introduce you to the following basic techniques of the Robot Boot Camp shown in Table I.1.

Table I.1   The Boot Camp Notes

Techniques	Description
Robot motion planning & programming	Arm movement
	Gripper programming
	End-effector movement
	Robot navigation
Programming the robot to use different types of sensors	Infrared sensors
	Ultrasonic sensors
	Touch sensors
	Light sensors
	RFID sensors
	Camera sensors
	Temperature sensors
	Sound sensors
	Analysis sensors
Motor use	Motors used in robot navigations
	Motors used in robotic arms, grippers, and end-effectors
	Motors used in sensor positioning
Decision-making	Robot action selection
	Robot direction selection
	Robot path selection
Instruction translation	Translating English instructions and commands into a programming language or instructional format that a robot can process

These techniques are the core techniques necessary to get a robot to execute almost any assigned task. Make note of these five areas because they represent the second step in building a solid foundation for robot programming.

BURT—Basic Universal Robot Translator

In this book, we use two aids to present the robot programs and common robot programming issues in an easy-to-understand and quick reference format. The first aid, *BURT (Basic Universal Robot*

Translator), is used to present all the code snippets, commands, and robot programs in this book. BURT shows two versions of each code snippet, command, or robot program:

- Plain English version

- Robot language version

BURT is used to translate from a simple, easy-to-understand English version of a set of instructions to the robot language version of those instructions.

In some cases the English version is translated into diagrams that represent the robot instructions. In other cases, BURT translates the English into standard programming languages like Java or C++. BURT can also be used to translate English instructions into robot visual instruction environments like Labview or LEGO's G language for Mindstorms robots.

The BURT Translations are numbered and can be used for quick reference guides on programming techniques, robot instructions, or commands. BURT Translations have two components; an input and an output component. The input component will contain the pseudocode, or RSVPs. The output component will contain the program listing, whether it be a standard language or visual instruction. They will be accompanied with the BURT Translation Input or Output logo as shown in Figure I.2.

In addition to BURT Translations, this book contains BURT Gotchas, a.k.a. BURT's **G**lossary of **T**echnical **C**oncepts and **H**elpful **A**cronyms. The world of robot programming is full of technical terms and acronyms that may be unfamiliar or tricky to recall. BURT Gotchas provide a convenient place to look up any acronym or some of the more technical terms used in this book. In some cases BURT Gotchas are listed at the end of the chapter in which they are first used, but a complete list of all of BURT Gotchas can be found in the book's glossary.

Figure I.2
BURT Translation
Input and Output
logos

BURT'S INPUT AND OUTPUT TRANSLATION LOGOS

BRON—Bluetooth Robot Oriented Network

The second aid is *BRON (Bluetooth Robot Oriented Network)*. We have put together a small team of robots that are connected and communicate through Bluetooth wireless protocols and the Internet. It is the responsibility of this team of robots to locate and retrieve useful tips, tricks, little-known facts, interviews, and news from the world of robot programming that the reader will find interesting and helpful. This material is presented in sections titled *BRON's Believe It or Not* and are identified by the logo shown in Figure I.3.

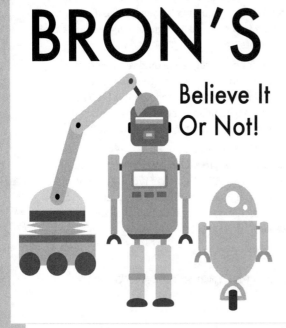

Figure I.3
BRON's Believe It or Not logo

These sections contain supplementary material that the reader can skip, but often offer additional insight on some idea that has been presented in the chapter. In some instances, a BRON's Believe It or Not contains news that is hot off the presses and relevant to some aspect of robot programming. In other instances, a BRON section contains excerpts from interviews of individuals making important contributions to the world of robotics or robot programming. In all cases, BRON's Believe It or Not sections are designed to give you a deeper understanding and appreciation for the world of robotics and robot programming.

Assumptions About the Reader's Robot(s)

Robot Programming: A Guide to Controlling Autonomous Robots can be read and much can be learned without any access to robots at all. Most chapters explain the concepts in plain English and are reinforced with diagrams. However, to get the maximum benefit from reading this book, it is assumed you will try out and test the commands, instructions, or programs on robots that you have access to.

We used and tested the instructions and programs in this book on several different types of robots, and the ideas presented in this book broadly apply to many classes of robots. If you have access to a robot with *at least one* capability from each column shown in Table I.2, you will be able to adapt any program in this book to your robot.

 note

We do show you how to program a robot to use other sensors beyond those listed in Table I.2. But the main ideas in the book can be tried and tested with only those listed in Table I.2.

Table I.2   The Boot Camp's Matrix of Robot Capabilities

Movement Capability	Sensing	Actuating	Control
Wheels	Infrared	Gripper	ARM7 Microcontroller
Bipedal	Ultrasonic	Robot arm	ARM9 Microcontroller
Quadruped	Camera	Pusher	LEGO Mindstorms EV3 Microcontroller
Hexaped (etc.)	Heat		LEGO Mindstorms NXT Microcontroller
Aerial	Light		Arduino
	Color		ARM Cortex/Edison Processor
	Touch		

How Midamba Learned to Program a Robot

In this book, we tell a short story of how free-spirited, fun-loving Midamba found himself in a precarious predicament. As luck would have it, his only chance out of the predicament required that he learn how to program robots. Although Midamba had some basic experience programming a computer, he had very little knowledge of robots and no experience programming them. So throughout the book, we use Midamba's predicament and his robot programming triumph as an example. We walk you through the same basic lessons that Midamba learned and the steps he had to take to successfully program his first robot.

WHAT IS A ROBOT ANYWAY?

Robot Sensitivity Training Lesson #1: *All robots are machines, but not all machines are robots.*

Ask any 10 people what a robot is and you're bound to get at least 10 different opinions: radio controlled toy dogs, automated bank teller machines, remote-controlled battle-bots, self-operating vacuum cleaners, drones that fly unattended, voice-activated smartphones, battery-operated action figures, titanium-plated hydraulic-powered battle chassis exoskeletons, and the list goes on.

It might not be easy to define what a robot is, but we all know one when we see one, right? The rapid growth of software-controlled devices has blurred the line between automated devices and robots. Just because an appliance or device is software-controlled does not make that appliance or device a robot. And being automated or self-operated is not enough for a machine to be given the privileged status of robot.

Many remote-controlled, self-operated devices and machines are given the status of robot but don't make the cut. Table 1.1 shows a few of the wide-ranging and sometimes contradictory dictionary type definitions for robot.

Table 1.1 Wide-Ranging, Sometimes Contradictory Definitions for the term *Robot*

Source	Definition
Urban Dictionary	1: A mechanical apparatus designed to do the work of a man.
Wikipedia	A mechanical or virtual artificial agent, usually an electro-mechanical machine that is guided by a computer program or electronic circuitry.
Merriam-Websters Dictionary	A machine that looks like a human being and performs various complex acts (as walking or talking) of a human being; *also :* a similar but fictional machine whose lack of capacity for human emotions is often emphasized.
Encyclopaedic Britannica	Any automatically operated machine that replaces human effort, though it may not resemble human beings in appearance or perform functions in a humanlike manner.
Webopedia	A device that responds to sensory input.

The Seven Criteria of Defining a Robot

Before we can embark on our mission to program robots, we need a good definition for what makes a robot a robot. So when does a self-operating, software-controlled device qualify as a robot? At ASC (Advanced Software Construction, where the authors build smart engines for robots and softbots), we require a machine to meet the following seven criteria:

1. It must be capable of sensing its external and internal environments in one or more ways through the use of its programming.

2. Its reprogrammable behavior, actions, and control are the result of executing a programmed set of instructions.

3. It must be capable of affecting, interacting with, or operating on its external environment in one or more ways through its programming.

4. It must have its own power source.

5. It must have a language suitable for the representation of discrete instructions and data as well as support for programming.

6. Once initiated it must be capable of executing its programming without the need for external intervention (controversial).

7. It must be a nonliving machine.

Let's take a look at these criteria in more detail.

Criterion #1: Sensing the Environment

For a robot to be useful or effective it must have some way of sensing, measuring, evaluating, or monitoring its environment and situation. What senses a robot needs and how the robot uses those senses in its environment are determined by the task(s) the robot is expected to perform. A robot might need to identify objects in the environment, record or distinguish sounds, take measurements of materials encountered, locate or avoid objects by touch, and so on. Without the capability of sensing its environment and situation in some way it would be difficult for a robot to perform tasks. In addition to having some way to sense its environment and situation the robot has to have the capability of accepting instructions on how, when, where, and why to use its senses.

Criterion #2: Programmable Actions and Behavior

There must be some way to give a robot a set of instructions detailing:

- What actions to perform
- When to perform actions
- Where to perform actions
- Under what situations to perform actions
- How to perform actions

As we see throughout this book *programming a robot* amounts to giving a robot a set of instructions on what, when, where, why, and how to perform a set of actions.

Criterion #3: Change, Interact with, or Operate on Environment

For a robot to be useful it has to not only sense but also change its environment or situation in some way. In other words, a robot has to be capable of doing something to something, or doing something with something! A robot has to be capable of making a difference in its environment or situation or it cannot be useful. The process of taking an action or carrying out a task has to affect or operate on the environment or there wouldn't be any way of knowing whether the robot's actions were effective. A robot's actions change its environment, scenario, or situation in some measurable way, and that change should be a direct result of the set of instructions given to the robot.

Criterion #4: Power Source Required

One of the primary functions of a robot is to perform some kind of action. That action causes the robot to expend energy. That energy has to be derived from some kind of power source, whether it be battery, electrical, wind, water, solar, and so on. A robot can operate and perform action only as long as its power source supplies energy.

Criterion #5: A Language Suitable for Representing Instructions and Data

A robot has to be given a set of instructions that determine how, what, where, when, and under which situations or scenarios an action is to take place. Some instructions for action are hard-wired into the robot. These are instructions the robot always executes as long as it has an active power source and regardless of the situation the robot is in. This is part of the "machine" aspect of a robot.

One of the most important differences between a regular machine and a robot is that a robot can receive new instructions without having to be rebuilt or have its hardware changed and without having to be rewired. A robot has a language used to receive instructions and commands. The robot's language must be capable of representing commands and data used to describe the robot's environment, scenario, or situation. The robot's language facility must allow instruction to be given without the need for physical rewiring. That is, a robot must be reprogrammable through a set of instructions.

Criterion #6: Autonomy Without External Intervention

We take the hard line and a have a controversial position with this requirement for a robot. For our purposes, true robots are fully autonomous. This position is not shared among all roboticists. Figure 1.1 is from our robot boot camp and shows the two basic categories of robot operation.

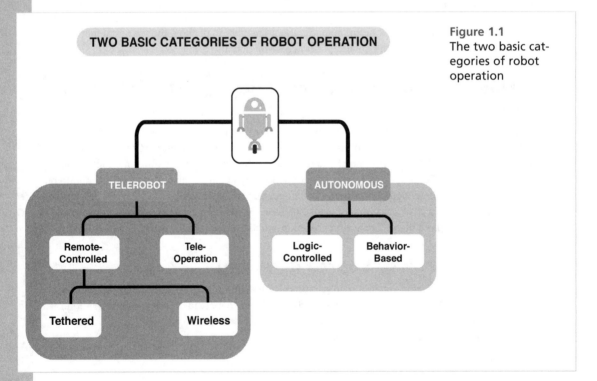

TWO BASIC CATEGORIES OF ROBOT OPERATION

TELEROBOT

Remote-Controlled

Tele-Operation

Tethered

Wireless

AUTONOMOUS

Logic-Controlled

Behavior-Based

Figure 1.1
The two basic categories of robot operation

Robot operation or robot control can be generally divided into two categories:

- Telerobots

- Autonomous robots

Like autonomous robots, telerobots do receive instructions, but their instructions are real-time and received in real-time or delayed-time from external sources (humans, computers, or other machines). Instructions are sent with a remote control of some type, and the robot performs the action the instruction requires. Sometimes one signal from a remote control triggers multiple actions once the robot receives the signal, and in other cases it is a one-for-one proposition—each signal corresponds to one action.

The key point to note here is that without the remote control or external intervention the robot does not perform any action. On the other hand, an autonomous robot's instructions are stored prior to the robot taking a set of actions. An autonomous robot has its marching orders and does not rely on remote controls to perform or initiate each action. To be clear, there are hybrid robots and variations on a theme where the telerobots have some autonomous actions and autonomous robots sometimes have a puppet mode.

But we make a distinction between fully autonomous and semiautonomous robots and throughout this book may refer to strong autonomy and weak autonomy. This book introduces the reader to the concepts and techniques involved in programming fully autonomous robots; remote-control programming techniques are not covered.

Criterion #7: A Nonliving Machine

Although plants and animals are sometimes considered programmable machines, we exclude them from the definition of robot. Many ethical considerations must be worked out as we build, program, and deploy robots. If we ever get to a point in robotics where the robots are considered living, then a change of the definition of robot will definitely be in order.

Robot Categories

While many types of machines may meet one or more of our seven criteria, to be considered a *true robot*, it is necessary that the machine meets at a minimum these seven criteria. To be clear, a robot may have more than these seven characteristics but not fewer. Fortunately, there is no requirement that a robot be humanlike, possess intelligence, or have emotions. In fact, most robots in use today have little in common with humans. Robots fall into three basic categories as shown in Figure 1.2.

The three categories of robot can also be further divided based on how they are operated and programmed. We previously described robots as being either telerobots or autonomous robots. So we can have ground-based, aerial, or underwater robots that are teleoperated or autonomous. Figure 1.3 shows a simple breakdown for aerial and underwater robots.

BASIC CATEGORIES FOR ROBOTS

Figure 1.2
Three basic categories of robots

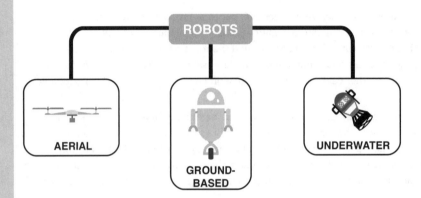

ROBOTS

AERIAL

GROUND-BASED

UNDERWATER

AERIAL AND UNDERWATER OPERATION MODES

Figure 1.3
The operation modes of aerial and under-water robots

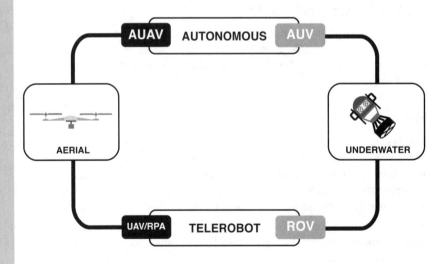

AUAV AUTONOMOUS AUV

AERIAL

UNDERWATER

UAV/RPA TELEROBOT ROV

AUAV   Autonomous Unmanned Aerial Vehicles
AUV   Autonomous Underwater Vehicles
UAV   Unmanned Aerial Vehicles
RPA   Remotely-Piloted Aircraft
ROV   Remotely-Operated Vehicles

Aerial and Underwater Robots

Aerial robots are referred to as UAVs (unmanned aerial vehicles) or AUAVs (autonomous unmanned aerial vehicles). Not every UAV and AUAV qualifies as a robot. Remember our seven criteria. Most UAVs are just machines, but some meet all seven criteria and can be programmed to perform tasks. Underwater robots are referred to as ROVs (remotely operated vehicles) and AUVs (autonomous underwater vehicles). Like UAVs, most ROVs are just machines and don't rise to the level of robot, but the ones that do can be programmed and controlled like any other robot.

As you might have guessed, aerial and underwater robots regularly face issues that ground-based robots typically don't face. For instance, underwater robots must be programmed to navigate and perform underwater and must handle all the challenges that come with aquatic environments (e.g., water pressure, currents, water, etc.). Most ground-based robots are not programmed to operate in aquatic environments and are not typically designed to withstand getting wet.

Aerial robots are designed to take off and land and typically operate hundreds or thousands of feet above ground in the air. UAV robot programming has to take into consideration all the challenges that an airborne machine must face (e.g., what happens if an aerial robot loses all power?). A ground-based robot may run out of power and simply stop.

UAVs and ROVs can face certain disaster if there are navigational or power problems. But ground-based robots can also sometimes meet disaster. They can fall off edges, tumble down stairs, run into liquids, or get caught in bad weather. Ordinarily robots are designed and programmed to operate in one of these categories. It is difficult to build and program a robot that can operate in multiple categories. UAVs don't usually do water and ROVs don't usually do air.

Although most of the examples in this book focus on ground-based robots, the robot programming concepts and techniques that we introduce can be applied to robots in all three categories. A robot is a robot. Figure 1.4 shows a simplified robot component skeleton.

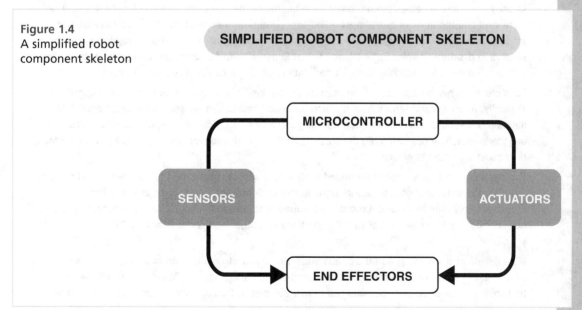

Figure 1.4
A simplified robot component skeleton

SIMPLIFIED ROBOT COMPONENT SKELETON

MICROCONTROLLER

SENSORS

ACTUATORS

END EFFECTORS

All true robots have the basic component skeleton shown in Figure 1.4. No matter which category a robot falls in (ground-based, aerial, or underwater), it contains at least four types of programmable components:

- One or more sensors

- One or more actuators

- One or more end-effectors/environmental effectors

- One or more microcontrollers

These four types of components are at the heart of most basic robot programming. In its simplest form, programming a robot boils down to programming the robot's sensors, actuators, and end-effectors using the controller. Yes, there is more to programming a robot than just dealing with the sensors, actuators, effectors, and controllers, but these components make up most of the stuff that is in, on, and part of the robot. The other major aspect of robot programming involves the robot's scenario, which we introduce later. But first let's take a closer (although simplified) look at these four basic programmable robot components.

What Is a Sensor?

Sensors are the robot's eyes and ears to the world. They are components that allow a robot to receive input, signals, data, or information about its immediate environment. Sensors are the robot's interface to the world. In other words, the sensors are the robot's senses.

Robot sensors come in many different types, shapes, and sizes. There are robot sensors that are sensitive to temperature; sensors that sense sound, infrared light, motion, radio waves, gas, and soil composition; and sensors that measure gravitational pull. Sensors can use a camera for vision and identify direction. Whereas human beings are limited to the five basic senses of sight, touch, smell, taste, and hearing, a robot has almost a limitless potential for sensors. A robot can have almost any kind of sensor and can be equipped with as many sensors as the power supply supports. We cover sensors in detail in Chapter 5, "A Close Look at Sensors." But for now just think of sensors as the devices that provide a robot with input and data about its immediate situation or scenario.

Each sensor is responsible for giving the robot some kind of feedback about its current environment. Typically, a robot reads values from a sensor or sends values to a sensor. But sensors don't just automatically sense. The robot's programming dictates when, how, where, and to what extent to use the sensor. The programming determines what mode the sensor is in and what to do with feedback received from the sensor.

For instance, say I have a robot equipped with a light sensor. Depending on the sensor's sophistication, I can instruct the robot to use its light sensor to determine whether an object is blue and then to retrieve only blue objects. If a robot is equipped with a sound sensor, I can instruct it to perform one action if it *hears* one kind of sound and execute another action if it *hears* a different kind of sound.

Not all sensors are created equal. For any sensor that you can equip a robot with, there is a range of low-end to high-end sensors in that category. For instance, some light sensors can detect only four colors, while other light sensors can detect 256 colors. Part of programming a robot includes

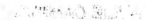

becoming familiar with the robot's sensor set, the sensor capabilities, and limitations. The more versatile and sophisticated a sensor is, the more elaborate the associated robot task can be.

A robot's effectiveness can be limited by its sensors. A robot that has a camera sensor with a range limited to a few inches cannot see something a foot away. A robot with a sensor that detects only light in the infrared cannot see ultraviolet light, and so on. We discuss robot effectiveness in this book and describe it in terms of the robot skeleton (refer to Figure 1.4). We rate a robot's potential effectiveness based on:

- Actuator effectiveness

- Sensor effectiveness

- End-effector effectiveness

- Microcontroller effectiveness

REQUIRE

By this simple measure of robot effectiveness, sensors count for approximately one-fourth of the potential robot effectiveness. We have developed a method of measuring robot potential that we call REQUIRE (Robot Effectiveness Quotient Used in Real Environments). We use REQUIRE as an initial litmus test in determining what we can program a robot to do and what we won't be able to program that robot to do. We explain REQUIRE later and use it as a robot performance metric throughout the book. It is important to note that 25% of a robot's potential effectiveness is determined by the quality of its sensors and how they can be programmed.

What Is an Actuator?

The actuator is the component(s) that provides movement of robot parts. Motors usually play the role of an actuator for a robot. The motor could be electric, hydraulic, pneumatic, or use some other source of potential energy. The actuator provides movement for a robot arm or movement for tractors, wheels, propellers, paddles, wings, or robotic legs. Actuators allow a robot to move its sensors and to rotate, shift, open, close, raise, lower, twist, and turn its components.

Programmable Robot Speed and Robot Strength

The actuators/motors ultimately determine how fast a robot can move. A robot's acceleration is tied to its actuators. The actuators are also responsible for how much a robot can lift or how much a robot can hold. Actuators are intimately involved in how much torque or force a robot can generate. Fully programming a robot involves giving the robot instructions for how, when, why, where, and to what extent to use the actuators. It is difficult to imagine or build a robot that has no movement of any sort. That movement might be external or internal, but it is present.

 tip

Recall criterion #3 from our list of robot requirements: It must be able to affect, interact with, or operate on its external environment in one or more ways through its programming.

It must be able to affect, interact with, or operate on its external environment in one or more ways through its programming.

It is very difficult to imagine or build a robot that has no movement of any sort. That movement might be external or internal but it is present.

A robot must operate on its environment in some way, and the actuator is a key component to this interaction. Like the sensor set, a robot's actuators can enable or limit its potential effectiveness. For example, an actuator that permits a robot to rotate its arm only 45 degrees would not be effective in a situation requiring a 90-degree rotation. Or if the actuator could move a propeller only 200 rpm where 1000 rpm is required, that actuator would prevent the robot from properly performing the required task.

The robot has to have the right type, range, speed, and degree of movement for the occasion. Actuators are the programmable components providing this movement. Actuators are involved with how much weight or mass a robot can move. If the task requires that the robot lift a 2-liter or 40-oz. container of liquid, but the robot's actuators only support a maximum of 12 ounces, the robot is doomed to fail.

A robot's effectiveness is often measured by how useful it is in the specified situation or scenario. The actuators often dictate how much work a robot can or cannot perform. Like the sensors, the robot's actuators don't simply actuate by themselves; they must be programmed. Like sensors, actuators range from low-end, simple functions to high-end, adaptable, sophisticated functions. The more flexible the programming facilities, the better. We discuss actuators in more detail in Chapter 7, "Programming Motors and Servos."

What Is an End-Effector?

The end-effector is the hardware that allows the robot to handle, manipulate, alter, or control objects in its environment. The end-effector is the hardware that causes a robot's actions to have an effect on its environment or scenario. Actuators and end-effectors are usually closely related. Most end-effectors require the use of or interaction with the actuators.

Robot arms, grippers, claws, and hands are common examples of end-effectors. End-effectors come in many shapes and sizes and have a variety of functions. For example, a robot can use components as diverse as drills, projectiles, pulleys, magnets, lasers, sonic blasts, or even fishnets for end-effectors. Like sensors and actuators, end-effectors are also under the control of the robot's programming.

 note

End-effectors also help to fulfill part of criterion #3 for our definition of robot. A robot has to be capable of doing something to something, or doing something with something! A robot has to make a difference in its environment or situation or it cannot be useful.

Part of the challenge of programming any robot is instructing the robot to use its end-effectors sufficiently to get the task done. The end-effectors can also limit a robot's successful completion of a task. Like sensors, a robot can be equipped with multiple effectors allowing it to manipulate

different types of objects, or similar objects simultaneously. Yes, in some cases the best end-effectors are the ones that work in groups of two, four, six, and so on. The end-effectors must be capable of interacting with the objects needed to be manipulated by the robot within the robot's scenario or environment. We take a detailed look at end-effectors in Chapter 7.

What Is a Controller?

The controller is the component that the robot uses to "control" its sensors, end-effectors, actuators, and movement. The controller is the "brain" of the robot. The controller function could be divided between multiple controllers in the robot but usually is implemented as a microcontroller. A microcontroller is a small, single computer on a chip. Figure 1.5 shows the basic components of a microcontroller.

Figure 1.5
The basic components of a micro-controller

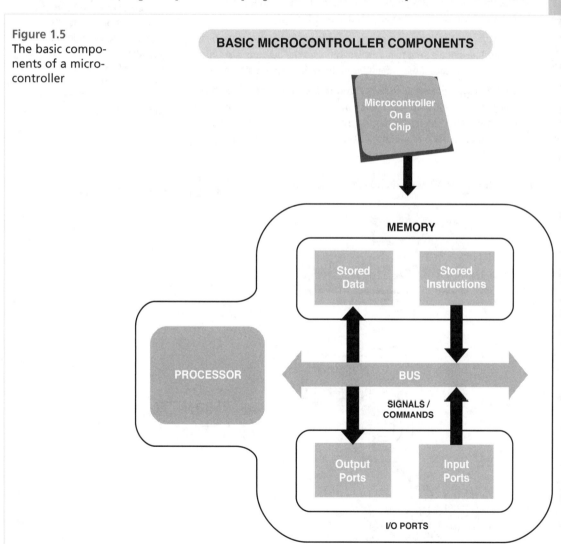

BASIC MICROCONTROLLER COMPONENTS

Microcontroller On a Chip

MEMORY

Stored Data

Stored Instructions

PROCESSOR

BUS

SIGNALS / COMMANDS

Output Ports

Input Ports

I/O PORTS

The controller or microcontroller is the component of the robot that is programmable and supports the programming of the robot's actions and behaviors. By definition, machines without at least one microcontroller can't be robots. But keep in mind being programmable is only one of seven criteria.

The controller controls and is the component that has the robot's memory. Note in Figure 1.5 there are four components. The processor is responsible for calculations, symbolic manipulation, executing instructions, and signal processing. The input ports receive signals from sensors and send those signals to the processor for processing. The processor sends signals or commands to the output ports connected to actuators so they can perform action.

The signals sent to sensors cause the sensors to be put into sensor mode. The signals sent to actuators initialize the actuators, set motor speeds, cause motor movement, stop motor movement, and so on.

So the processor gets feedback through the input ports from the sensors and sends signals and commands to the output ports that ultimately are directed to motors; robot arms; and robot-movable parts such as tractors, pulley wheels, and other end-effectors.

 note

The processor executes instructions. A processor has its own kind of machine language. A set of instructions sent to a processor must ultimately be translated into the language of the processor. So if we start out with a set of instructions in English we want to give a processor to execute, that set of instructions must ultimately be translated into instructions the processor inside the microcontroller can understand and execute. Figure 1.6 shows the basic steps from "bright idea to processor" machine instruction.

BRIGHT IDEA TO MACHINE LANGUAGE TRANSLATION

Figure 1.6
Bright idea to machine language

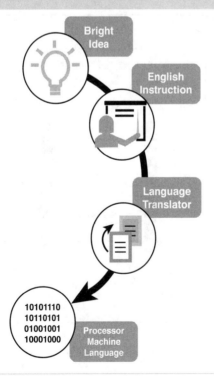

In this book, BURT, discussed previously in the robot boot camp, is used to translate bright ideas to machine language, so you can be clear on what takes place during the robot programming process. Unless you have a microcontroller that uses English or some other natural language as its internal language, all instructions, commands, and signals have to be translated from whatever form they start out in to a format that can be recognized by the processor within the microcontroller. Figure 1.7 shows a simple look at the interaction between the sensors, actuators, end-effectors, and the microcontroller.

Figure 1.7
The basic interaction between the microcontroller, sensors, actuators, and end-effectors

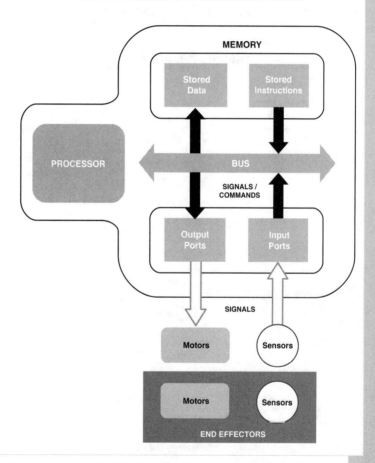

The memory component in Figure 1.7 is the place where the robot's instructions are stored and where the robot's current operation data is stored. The current operation data includes data from the sensors, actuators, processor, or any other peripherals needed to store data or information that must be ultimately processed by the controller's processor. We have in essence reduced a robot to its basic robot skeleton:

- Sensors
- Actuators

■ End-effectors

■ Microcontrollers

At its most basic level, programming a robot amounts to manipulating the robot's sensors, end-effectors, and motion through a set of instructions given to the microcontroller. Regardless of whether we are programming a ground-based, aerial, or underwater robot, the basic robot skeleton is the same, and a core set of primary programming activities must be tackled. Programming a robot to autonomously execute a task requires we communicate that task in some way to the robot's microcontroller. Only after the robot's microcontroller has the task can it be executed by the robot. Figure 1.8 shows our translated robot skeleton.

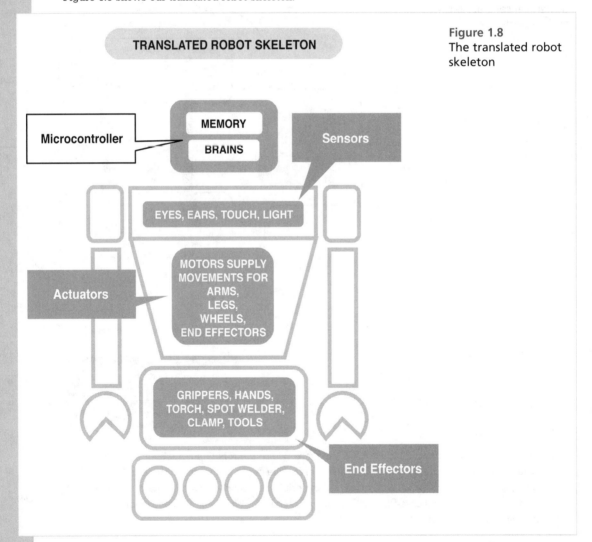

Figure 1.8
The translated robot skeleton

Figure 1.8 shows what the basic robot components are analogous to. Robot sensors, actuators, and end-effectors can definitely come in more elaborate forms, but Figure 1.8 shows some of the more commonly used components and conveys the basic ideas we use in this book.

 note

We place a special emphasis on the microcontroller because it is the primary programmable component of a robot, and when robot programming is discussed, the microcontroller is usually what is under discussion. The end-effectors and sensors are important, but the microcontroller is in the driver's seat and is what sets and reads the sensors and controls the movement of the robots and the use of the end-effectors. Table 1.2 lists some of the commonly used microcontrollers for low-cost robots.

Table 1.2 Commonly Used Microcontrollers

Microcontroller	Robot Platform
Atmega328	Arduino Uno
Quark	Intel Edison
ARM7, ARM9	RS Media, Mindstorms NXT, EV3
ARM Cortex	Vex
CM5/ATmega 128	Bioloid

Although most of examples in this book were developed using Atmega, ARM7, and ARM9 microcontrollers, the programming concepts we introduce can be applied to any robot that has the basic robot skeleton referred to in Figure 1.4 and that meets our seven robot criteria.

What Scenario Is the Robot In?

The robot skeleton tells only half the story of programming a robot. The other half, which is not part of the actual robot, is the robot's scenario or situation. Robots perform tasks of some type in a particular context. For a robot to be useful it must cause some kind of effect in context. The robot's tasks and environments are not just random, unspecified notions. Useful robots execute tasks within specified scenarios or situations. The robot is given some role to play in the scenario and situation. Let's take, for example, a robot at a birthday party in Figure 1.9.

We want to assign the robot, let's call him BR-1, two tasks:

- Light the candles on the cake.
- Clear away the dishes and cups after the party ends.

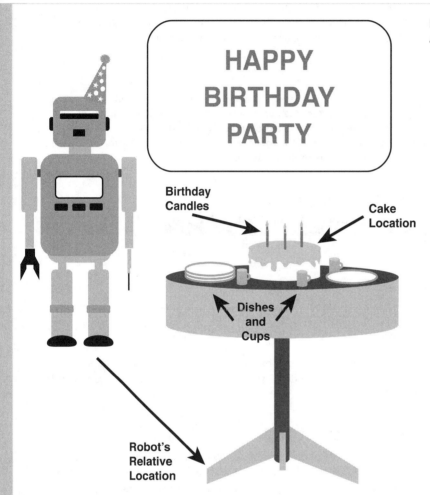

Figure 1.9
A birthday party robot

HAPPY
BIRTHDAY
PARTY

**Birthday
Candles**

**Cake
Location**

**Dishes
and
Cups**

**Robot's
Relative
Location**

The birthday party is the robot's *scenario*. The robot, BR-1, has the role of lighting the candles and clearing away the dishes and cups. Scenarios come with expectations. Useful robots are expectation driven. There are expectations at a birthday party. A typical birthday party has a location, cake, ice cream, guests, someone celebrating a birthday, a time, and possibly presents. For BR-1 to carry out its role at the birthday party, it must have instructions that address the scenario or situation-specific things like

- The location of the cake
- How many candles to light
- The robot's location relative to the cake
- When to light the candles

- How and when the party ends

- The number of dishes and cups and so on

The robot's usefulness and success are determined by how well it fulfills its role in the specified situation. Every scenario or situation has a location, a set of objects, conditions, and a sequence of events. Autonomous robots are situated within scenarios and are expectation driven. When a robot is programmed, it is expected to participate in and affect a situation or scenario in some way. Representing and communicating the scenario and set of expectations to the robot is the second half of the story of robot programming.

 tip

In a nutshell, programming a robot to be useful amounts to programming the robot to use its sensors and end-effectors to accomplish its role and meet its expectations by executing a set of tasks in some specified situation or scenario.

Programming a useful robot can be broken down into two basic areas:

- Instructing the robot to use its basic capabilities to fulfill some set of expectations

- Explaining to the robot what the expectations are in some given situation or scenario

An autonomous robot is useful when it meets its expectations in the specified situation or scenario through its programming without the need for human intervention. So then half the job requires programming the robot to execute some task or set of tasks.

The other half requires instructing the robot to execute its function in the specified scenario or scenarios. Our approach to robot programming is scenario driven. Robots have roles in situations and scenarios, and those situations and scenarios must be part of the robot's instructions and programming for the robot to be successful in executing its task(s).

Giving the Robot Instructions

If we expect a robot to play some role in some scenario, how do we tell it what to do? How do we give it instructions? In the answers to these questions lie the adventure, challenge, wonder, dread, and possibly regret in programming a robot.

Humans use natural languages, gestures, body language, and facial expressions to communicate. Robots are machines and only understand the machine language of the microcontroller. Therein lies the rub. We speak and communicate one way and robots communicate another. And we do not yet know how to build robots to directly understand and process human communication. So even if we have a robot that has the sensors, end effectors, and capabilities to do what we require it to do, how do we communicate the task? How do we give it the right set of instructions?

Every Robot Has a Language

So exactly what language does a robot understand? The native language of the robot is the language of the robot's microcontroller. No matter how one communicates to a robot, ultimately that communication has to be translated into the language of the microcontroller. The microcontroller is a computer, and most computers speak machine language.

Machine Language

Machine language is made of 0s and 1s. So, technically, the only language a robot really understands is strings and sequences of 0s and 1s. For example:

0000001, 1010101, 00010010, 10101010, 11111111

And if you wanted to (or were forced to) write a set of instructions for the robot in the robot's native language, it would consist of line upon line of 0s and 1s. For example, Listing 1.1 is a simple ARM machine language (sometimes referred to as binary language) program.

Listing 1.1 ARM Machine Language Program

```
1110010110011111000100000010000
1110010110011111000100000001000
1110000010000001010100000000000
1110010110001111010100000001000
```

This program takes numbers from two memory locations in the processor, adds them together, and stores the result in a third memory location. Most robot controllers speak a language similar to the machine language shown in Listing 1.1.

The arrangement and number of 0s and 1s may change from controller to controller, but what you see is what you get. Pure machine language can be difficult to read, and it's easy to flip a 1 or 0, or count the 1s and 0s incorrectly.

Assembly Language

Assembly language is a more readable form of machine language that uses short symbols for instructions and represents binary using hexadecimal or octal notation. Listing 1.2 is an assembly language example of the kind of operations shown in Listing 1.1.

Listing 1.2 Assembly Language Version of Listing 1.1

```
LDR R1, X
LDR R0, Y
ADD R5, R1, R0
STR R5, Z
```

While Listing 1.2 is more readable than Listing 1.1 and although entering assembly language programs is less prone to errors than entering machine language programs, the microcontroller assembly language is still a bit cryptic. It's not exactly clear we're taking two numbers X and Y, adding them together, and then storing the result in Z.

 note

Machine language is sometimes referred to as a *first-generation language*. Assembly language is sometimes referred to as a *second-generation language*.

In general, the closer a computer language gets to a natural language the higher the generation designation it receives. So a third-generation language is closer to English than a second-generation language, and a fourth-generation language is closer than a third-generation language, and so on.

So ideally, we want to use a language as close to our own language as possible to instruct our robot. Unfortunately, a higher generation of language typically requires more hardware resources (e.g., circuits, memory, processor capability) and requires the controller to be more complex and less efficient. So the microcontrollers tend to have only second-generation instruction sets.

What we need is a universal translator, something that allows us to write our instructions in a human language like English or Japanese and automatically translate it into machine language or assembly language. The field of computers has not yet produced such a universal translator, but we are halfway there.

Meeting the Robot's Language Halfway

Compilers and interpreters are software programs that convert one language into another language. They allow a programmer to write a set of instructions in a higher generation language and then convert it to a lower generation language. For example, the assembly language program from Listing 1.2

```
LDR R1, X
LDR R0, Y
ADD R5, R1, R0
STR R5, Z
```

can be written using the instruction

```
Z = X + Y
```

Notice in Figure 1.10 that the compiler or interpreter converts our higher-level instruction into assembly language, but an assembler is the program that converts the assembly language to machine language. Figure 1.10 also shows a simple version of the idea of a tool-chain.

Tool-chains are used in the programming of robots. Although we can't use natural languages yet, we have come a long way from assembly language as the only choice for programming robots. In fact, today we have many higher-level languages for programming robots, including graphic languages such as Labview, graphic environments such as Choreograph, puppet modes, and third-, fourth-, and fifth-generation programming languages.

Figure 1.11 shows a taxonomy of some commonly used generations of robot programming languages.

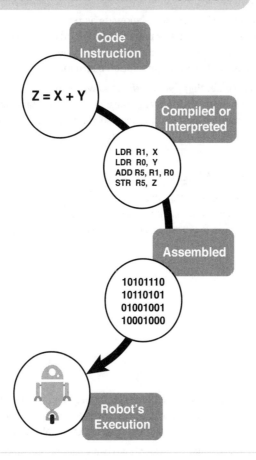

COMPILER AND INTERPRETER TRANSLATION

Code Instruction

$Z = X + Y$

Compiled or Interpreted

```
LDR  R1, X
LDR  R0, Y
ADD R5, R1, R0
STR  R5, Z
```

Assembled

```
10101110
10110101
01001001
10001000
```

Robot's Execution

Figure 1.10
Compiler and interpreter translations

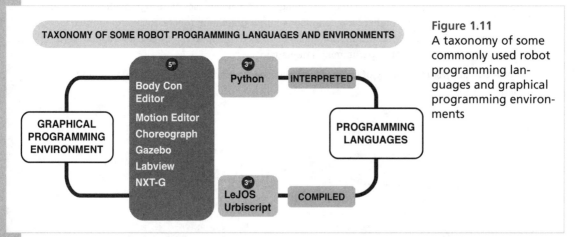

TAXONOMY OF SOME ROBOT PROGRAMMING LANGUAGES AND ENVIRONMENTS

5th
Body Con Editor
Motion Editor
Choreograph
Gazebo
Labview
NXT-G

3rd
Python   INTERPRETED

3rd
LeJOS
Urbiscript   COMPILED

GRAPHICAL PROGRAMMING ENVIRONMENT

PROGRAMMING LANGUAGES

Figure 1.11
A taxonomy of some commonly used robot programming languages and graphical programming environments

Graphic languages are sometimes referred to as fifth- and sixth-generation programming languages. This is in part because graphic languages allow you to express your ideas more naturally rather than expressing your ideas from the machine's point of view. So one of the challenges of instructing or programming robots lies in the gap between expressing the set of instructions the way you think of them versus expressing the set of instructions the way a microcontroller sees them.

Graphical robot programming environments and graphic languages attempt to address the gap by allowing you to program the robots using graphics and pictures. The Bioloid[1] Motion Editor and the Robosapien[2] RS Media Body Con Editor are two examples of this kind of environment and are shown in Figure 1.11.

These kinds of environments work by allowing you to graphically manipulate or set the movements of the robot and initial values of the sensors and speeds and forces of the actuators. The robot is programmed by setting values and moving graphical levers and graphical controls. In some cases it's a simple matter of filling out a form in the software that has the information and data that needs to be passed to the robot. The graphical environment isolates the programmer from the real work of programming the robot.

It is important to keep in mind that ultimately the robot's microcontroller is expecting assembly/machine language. Somebody or something has to provide the native instructions. So these graphical environments have internal compilers and interpreters doing the work of converting the graphical ideas into lower-level languages and ultimately into the controller's machine language. Expensive robotic systems have had these graphical environments for some time, but they are now available for low-cost robots. Table 1.3 lists several commonly used graphical environments/languages for low-cost robotic systems.

Table 1.3 Examples of Graphical Robot Programming Environments

Environment	Robot System	Simulation Possible?	Puppet Modes?
Body Con Editor	Robosapien RS Media	Yes	Yes
Motion Editor	Bioloid	Yes	Yes
Choreograph	Nao	Yes	Yes
Gazebo	OSRF	Yes	No
Labview	Mindstorms EV3, NXT	Yes	No

Puppet Modes

Closely related to the idea of visually programming the robot is the notion of direct manipulation or *puppet mode*. Puppet modes allow the programmer to manipulate a graphic of the robot using the keyboard, mouse, touchpad, touchscreen, or some other pointing device, as well as allowing the programming to physically manipulate the parts of the robot to the desired positions.

The puppet mode acts as a kind of action recorder. If you want the robot's head to turn left, you can point the graphic of the robot's head to the left and the puppet mode remembers that. If you want the robot to move forward, you move the robot's parts into the forward position, whether it be legs, wheels,

---

1 Bioloid is a modular robotics system manufactured by Robotis.
2 Robosapien is a self-contained robotics system manufactured by Wowee.

tractors, and so on, and the puppet mode remembers that. Once you have recorded all the movements you want the robot to perform, you then send that information to the robot (usually using some kind of cable or wireless connection), and the robot physically performs the actions recorded in puppet mode.

Likewise, if the programmer puts the robot into puppet mode (provided the robot has a puppet mode), then the physical manipulation of the robot that the programmer performs is recorded. Once the programmer takes the robot out of record mode, the robot executes the sequence of actions recorded during the puppet mode session. Puppet mode is a kind of programming by imitation. The robot records how it is being manipulated, remembers those manipulations, and then duplicates the sequence of actions. Using a puppet mode allows the programmer to bypass typing in a sequence of instructions and relieves the programmer of having to figure out how to represent a sequence of actions in a robot language. The availability of visual languages, visual environments, and puppet modes for robot systems will only increase, and their sophistication will increase also. However, most visual programming environments have a couple of critical shortcomings.

How Is the Robot Scenario Represented in Visual Programming Environments?

Recall that half the effort of programming an autonomous robot to be effective requires instructing the robot on how to execute its role within a scenario or situation. Typically, visual robot programming environments such as those listed in Table 1.3 have no easy way to represent the robot's situation or scenario. They usually only include a simulation of a robot in a generic space with a potential simple object.

For example, there would be no easy way to represent our birthday party scenario in the average visual robot environment. The visual/graphical environments and puppet mode approach are effective at programming a sequence of movements that don't have to interact with anything. But autonomous robots need to affect the environment to be useful and need to be instructed about the environment to be effective, and graphical environments fall short in this regard. The robot programming we do in this book requires a more effective approach.

Midamba's Predicament

Robots have a language. Robots can be instructed to execute tasks autonomously. Robots can be instructed about situations and scenarios. Robots can play roles within situations and scenarios and effect change in their environments. This is the central theme of this book: *instructing a robot to execute a task in the context of a specific situation, scenario, or event.* And this brings us to the unfortunate dilemma of our poor stranded Sea-Dooer, Midamba.

Robot Scenario #1

When we last saw Midamba in our robot boot camp in the introduction, his electric-powered Sea-Doo had run low on battery power. Midamba had a spare battery, but the spare had acid corrosion on the terminals. He had just enough power to get to a nearby island where he might possibly find help.

Unfortunately for Midamba, the only thing on the island was an experimental chemical facility totally controlled by autonomous robots. But all was not lost; Midamba figured if there was a chemical in the facility that could neutralize the battery acid, he could clean his spare battery and be on his way.

A front office to the facility was occupied by a few robots. There were also some containers, beakers, and test tubes of chemicals, but he had no way to determine whether they could be used. The office was separated from a warehouse area where other chemicals were stored, and there was no apparent way into the area. Midamba could see the robots from two monitors where robots were moving about in the warehouse area transporting containers, marking containers, lifting objects, and so on.

In the front office was also a computer, a microphone, and a manual titled *Robot Programming: A Guide to Controlling Autonomous Robots* by Cameron Hughes and Tracey Hughes. Perhaps with a little luck he could find something in the manual that would instruct him on how to program one of the robots to find and retrieve the chemical he needed. Figure 1.12 and Figure 1.13 shows the beginning of Midamba's Predicament.

Figure 1.12
Midamba's predicament

Figure 1.13
Midamba's predicament continues.

So we follow Midamba as an example, as he steps through the manual. The first chapter he turns to is "Robot Vocabularies."

What's Ahead?

In Chapter 2, "Robot Vocabularies," we will discuss the language of the robot and how human language will be translated into the language the robot can understand in order for it to perform the tasks you want it to perform autonomously.

2

ROBOT VOCABULARIES

Robot Sensitivity Training Lesson #2: *A robot's actions are only as good as the instructions given that describe those actions.*

Robots have language. They speak the language of the microcontroller. Humans speak what are called natural languages (e.g., Cantonese, Yoruba, Spanish). We communicate with each other using natural languages, but to communicate to robots we have to either build robots that understand natural languages or find some way to express our intentions for the robot in a language it can process.

At this point in time little progress has been made in building robots that can fully understand natural languages. So we are left with the task of finding ways to express our instructions and intentions in something other than natural language.

Recall that the role of the interpreter and compiler (shown previously in Figure 1.10 and here again in Figure 2.1) is to translate from a higher-level language like Java or C++ to a lower-level language like assembly, byte-code, or machine language (binary),

 note

Important Terminology: The controller or microcontroller is the component of the robot that is programmable and supports the programming of the robot's actions and behaviors. By definition, machines without at least one microcontroller are not robots.

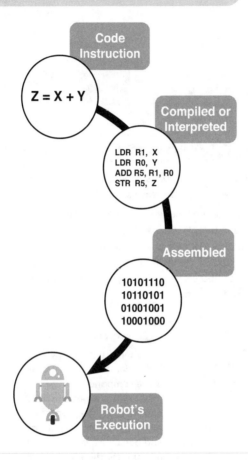

COMPILER AND INTERPRETER TRANSLATION

Code
Instruction

Z = X + Y

Compiled or
Interpreted

LDR R1, X
LDR R0, Y
ADD R5, R1, R0
STR R5, Z

Assembled

10101110
10110101
01001001
10001000

Robot's
Execution

Figure 2.1
The role of the interpreter and
compiler translation from a higher-
level language to a lower-level
language

One strategy is to meet the robot halfway. That is, find a language easy for humans to use and not
too hard to translate into the language of the robot (i.e., microcontroller), and then use a compiler
or interpreter to do the translation for us. Java and C++ are high-level languages used to program
robots. They are third-generation languages and are a big improvement over programming directly
in machine language or assembly language (second generation), but they are not natural languages.
It still requires additional effort to express human ideas and intentions using them.

Why the Additional Effort?

It usually takes several robot instructions to implement *one* high-level or human language instruction. For example, consider the following simple and single human language instruction:

Robot, hold this can of oil

This involves several instructions on the robot side. Figure 2.2 shows a partial BURT (Basic Universal Robot Translator) translation of this instruction to Arduino sketch code (C Language) that could be used to communicate this instruction to a robot and RS Media—a bipedal robot from WowWee that uses an ARM9 microcontroller with embedded Linux—code (Java Language).

Figure 2.2
A partial BURT translation of a natural language instruction translated to RS Media code (Java Language) and to Arduino sketch code (C Language).

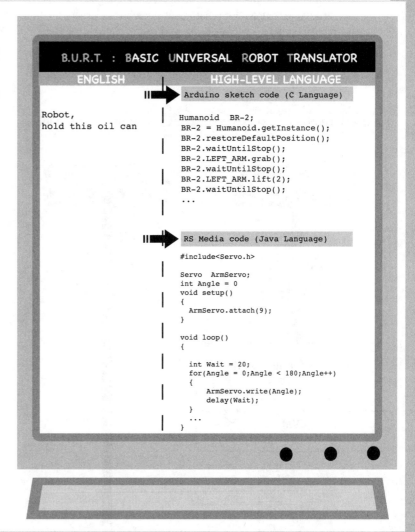

B.U.R.T. : BASIC UNIVERSAL ROBOT TRANSLATOR

ENGLISH | HIGH-LEVEL LANGUAGE

Arduino sketch code (C Language)

Robot,
hold this oil can

```
Humanoid  BR-2;
BR-2 = Humanoid.getInstance();
BR-2.restoreDefaultPosition();
BR-2.waitUntilStop();
BR-2.LEFT_ARM.grab();
BR-2.waitUntilStop();
BR-2.LEFT_ARM.lift(2);
BR-2.waitUntilStop();
...
```

RS Media code (Java Language)

```
#include<Servo.h>

Servo  ArmServo;
int Angle = 0
void setup()
{
   ArmServo.attach(9);
}

void loop()
{
   int Wait = 20;
   for(Angle = 0;Angle < 180;Angle++)
   {
      ArmServo.write(Angle);
      delay(Wait);
   }
   ...
}
```

Notice the BURT translation requires multiple lines of code. In both cases, a compiler is used to translate from a high-level language to ARM assembly language of the robot. The Java code is further removed from ARM than the Arduino language. The Java compiler compiles the Java into machine independent byte-code, which requires another layer to translate it into a machine-specific code. Figure 2.3 shows some of the layers involved in the translation of C and Java programs to ARM assembly.

Many of the microcontrollers used to control low-cost robots are ARM-based microcontrollers. If you are interested in using or understanding assembly language to program a low-cost robot, the ARM controller is a good place to start.

Notice in Figure 2.3 the software layers required to go from C or Java to ARM assembly. But what about the work necessary for the translation between *"robot hold this can of oil"* and the C or Java code? The BURT translator shows us that we start out with an English instruction and somehow we translate that into C and Java.

note

Important Terminology: Assembly language is a more readable form of machine language that uses short symbols for instructions and represents binary using hexadecimal or octal notation.

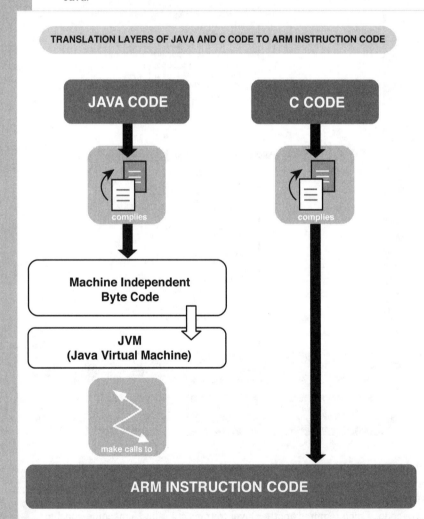

TRANSLATION LAYERS OF JAVA AND C CODE TO ARM INSTRUCTION CODE

JAVA CODE

C CODE

complies

complies

Machine Independent
Byte Code

JVM
(Java Virtual Machine)

make calls to

ARM INSTRUCTION CODE

Figure 2.3
Some of the layers involved in the translation of C and Java programs to ARM assembly

How is this done? Why is this done? Notice the instructions in the BURT translation do not mention *can* or *oil* or the concept of "hold this." These concepts are not directly part of the microcontroller language, and they are not built-in commands in the high-level C or Java languages.

They represent part of a robot vocabulary we want to use and we want the robot to understand (but does not yet exist).

However, before we can decide what our robot vocabulary will look like we need to be familiar with the robot's profile and its capabilities. Consider these questions:

- What kind of robot is it?

- What kind and how many sensors does the robot have?

- What kind and how many end-effectors does the robot have?

- How does it move and so on?

The robot vocabulary has to be tied to the robot's capabilities. For example, we cannot give a robot an instruction to lift something if the robot has no hardware designed for lifting. We cannot give a robot an instruction to move to a certain location if the robot is not mobile.

We can approach the robot and its vocabulary in two ways:

- Create a vocabulary and then obtain a robot that has the related capabilities.

- Create a vocabulary based on the capabilities of the robot that will be used.

In either case, the robot's capabilities must be identified. A robot capability matrix can be used for this purpose.

Table 2.1 shows a sample capability matrix for the robot called Unit2, the RS Media robot we will use for some of the examples in this book.

 note

A *robot vocabulary* is the language *you* use to assign a robot tasks for a specific situation or scenario. A primary function of programming useful autonomous robots is the creation of robot vocabularies.

 note

Important Terminology: Sensors are the robot's eyes and ears. They are components that allow a robot to receive input, signals, data, or information about its immediate environment. Sensors are the robot's interface to the world.

The end-effector is the hardware that allows the robot to handle, manipulate, alter, or control objects in its environment. The end-effector is the hardware that causes a robot's actions to have an effect on its environment.

Table 2.1 Capability Matrix for the RS Media Robot

Robot Name	Micro-controller	Sensors	End-effectors	Mobility	Communication
Unit2	ARM7 (Java Language)	3-color light Camera Sound Touch/bump	Right arm gripper Left arm gripper	Biped	USB port

A capability matrix gives us an easy-to-use reference for the robot's potential skill set. The initial robot vocabulary has to rely on the robot's base set of capabilities.

You are free to design your own robot vocabularies based on the particular robot capability and scenarios your robot will be used in. Notice in Table 2.1 that the robot Unit2 has a camera sensor and color sensors. This allows Unit2 to see objects. This means our robot vocabulary could use words like *watch*, *look*, *photograph*, or *scan*.

Being a biped means that Unit2 is mobile. Therefore, the vocabulary could include words like move, travel, walk, or proceed. The fact that Unit2 has a USB port suggests our robot vocabulary could use words like *open*, *connect*, *disconnect*, *send*, and *receive*. The initial robot vocabulary forms the basis of the instructions you give to the robot causing it to perform the required tasks.

 note

There is no one right robot vocabulary. Sure, there are some common actions that robots use, but the vocabulary chosen for a robot in any given situation is totally up to the designer or programmer.

If we look again at the capability matrix in Table 2.1, a possible vocabulary that would allow us to give Unit2 instructions might include

- Travel forward
- Scan for a blue can of oil
- Lift the blue can of oil
- Photograph the can of oil

One of the most important concepts you are introduced to in this book is how to come up with a robot vocabulary, how to implement a vocabulary using the capabilities your robot has, and how to determine the vocabularies the robot's microcontroller can process. In the preceding example, the Unit2's ARM9 microcontroller would not recognize the following command:

- Travel forward or scan for a blue can of oil

But that is the vocabulary we want to use with the robot. In this case, the tasks that we want the robot to perform involve traveling forward, looking for a blue can of oil, lifting it up, and then photographing it. We want to be able to communicate this set of tasks as simply as we can.

We have some situation or scenario we want an autonomous robot to play a role in. We develop a list of instructions that capture or describe the role we want the robot to play, and we want that list of instructions to be representative of the robot's task. Creating a robot vocabulary is part of the overall process of successfully programming an autonomous robot that performs useful roles in particular situations or scenarios that need a vocabulary.

Identify the Actions

One of the first steps in coming up with a robot vocabulary is to create the capability matrix and then based on the matrix identify the kinds of actions the robot can perform. For example, in the case of Unit2 from Table 2.1, the actions might be listed as

- Scan
- Lift

- Pick up

- Travel

- Stop

- Connect

- Disconnect

- Put Down

- Drop

- Move Forward

- Move Backward

Eventually we have to instruct the robot what we mean by scan, travel, connect, and so on. We described Unit2 as potentially being capable of scanning for a blue can of oil. Where does *"blue can of oil" fit* into our basic vocabulary?

Although Table 2.1 specifies our robot has color sensors, there is nothing about cans of oil in the capability matrix. This brings us to another important point that we expand on throughout the book:

- Half the robot's vocabulary is based on the robot's profile or capabilities.

- The other half of the robot's vocabulary is based on the scenario or situation where the robot is expected to operate.

These are important concepts essential to programming useful autonomous robots. The remainder of this book is spent elaborating on these two important ideas.

The Autonomous Robot's ROLL Model

Designing and implementing how you are going to describe your robot to use its basic capabilities, and designing and implementing how you are going to describe to your robot its role in a given situation or scenario is what programming autonomous robots is all about. To program your robot you need to be able to describe

- What to do

- Whon to do it

- Where to do it

- How to do it

Of equal importance is describing to your robot what the "it" is, or what does "where" or "when" refer to. As we see, this requires several levels of robot vocabulary.

For our purposes, an autonomous robot has seven levels of robot vocabulary that we call the robot's ROLL (Robot Ontology Language Level) model. Figure 2.4 shows the seven levels of the robot's ROLL model.

 note

In this book, we use the word *ontology* to refer to a description of a robot scenario or robot situation.

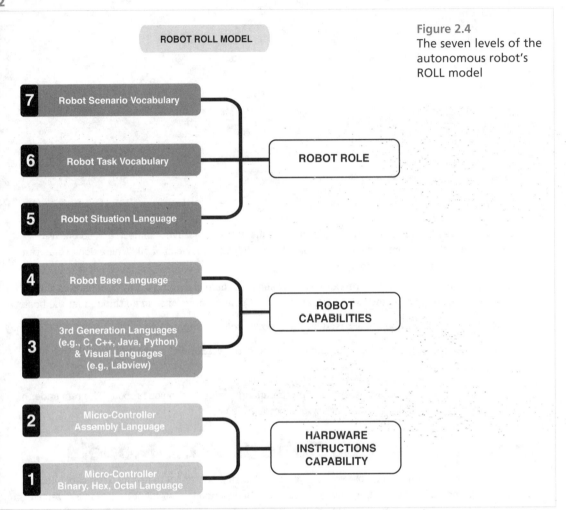

ROBOT ROLL MODEL

7 Robot Scenario Vocabulary

6 Robot Task Vocabulary

5 Robot Situation Language

ROBOT ROLE

4 Robot Base Language

3 3rd Generation Languages (e.g., C, C++, Java, Python) & Visual Languages (e.g., Labview)

ROBOT CAPABILITIES

2 Micro-Controller Assembly Language

1 Micro-Controller Binary, Hex, Octal Language

HARDWARE INSTRUCTIONS CAPABILITY

Figure 2.4
The seven levels of the autonomous robot's ROLL model

For now, we focus only on the robot programming that can occur at any one of the seven levels. The seven levels can be roughly divided into two groups:

- Robot capabilities
- Robot roles

 tip

Keep the ROLL model handy because we refer to it often.

Robot Capabilities

The robot vocabulary in levels 1 through 4 is basically directed at the capabilities of the robot—that is, what actions the robot can take and how those actions are implemented for the robot's hardware. Figure 2.5 shows how the language levels are related.

Figure 2.5
The relationship between robot capability language levels

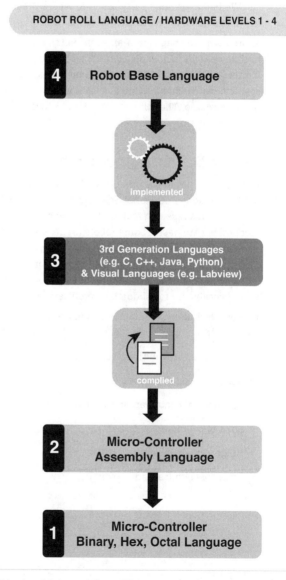

ROBOT ROLL LANGUAGE / HARDWARE LEVELS 1 - 4

4 Robot Base Language

implemented

3 3rd Generation Languages
(e.g. C, C++, Java, Python)
& Visual Languages (e.g. Labview)

complied

2 Micro-Controller
Assembly Language

1 Micro-Controller
Binary, Hex, Octal Language

The level 4 robot vocabulary is implemented by level 3 instructions. We expand on this later. We use level 3 instructions to define what we mean by the level 4 robot base vocabulary. As you recall, level 3 instructions (third-generation languages) are translated by interpreters or compilers into level 2 or level 1 microcontroller instructions.

42   Robot Vocabularies

2

Robot Roles in Scenarios and Situations

Levels 5 through 7 from the robot ROLL model shown in Figure 2.4 handle the robot vocabulary used to give the robot instructions about particular scenarios and situations. Autonomous robots execute tasks for particular scenarios or situations.

Let's take, for example, BR-1, our birthday robot from Chapter 1, "What Is a Robot Anyway?", Figure 1.9. We had a birthday party scenario where the robot's task was to light the candles and remove the plates and cups after the party was over.

A situation is a snapshot of an event in the scenario. For instance, one situation for BR-1, our birthday party robot, is a cake on a table with unlit candles. Another situation is the robot moving toward the cake. Another situation is positioning the lighter over the candles, and so on. Any scenario can be divided into a collection of situations. When you put all the situations together you have a scenario. The robot is considered successful if it properly executes all the tasks in its role for a given scenario.

Level 5 Situation Vocabularies

Level 5 has a vocabulary that describes particular situations in a scenario. Let's take a look at some of the situations from Robot Scenario #1 in Chapter 1, where Midamba finds himself stranded. There are robots in a research facility; there are chemicals in the research facility. Some chemicals are liquids; others are gases. Some robots are mobile and some are not. The research facility has a certain size, shelves, and containers. The robots are at their designated locations in the facility. Figure 2.6 shows some of the things that have to be defined using the level 5 vocabulary for Midamba's predicament.

One of the situations Midamba has in Scenario #1 requires a vocabulary the robots can process that handles the following:

- Size of their area in the research facility

- Robot's current location

- Locations of containers in the area

- Sizes of containers in the area

- Heights of shelves

- Types of chemicals and so on

 note

Notice that a robot vocabulary that describes these properties differs from the kind of vocabulary necessary to describe the robot's basic capabilities. Before each action a robot takes, we can describe the current situation. After each action the robot takes, we can describe the current situation. The actions a robot takes always change the situation in one or more ways.

Figure 2.6
Properties of situations
described by level 5 vocabulary

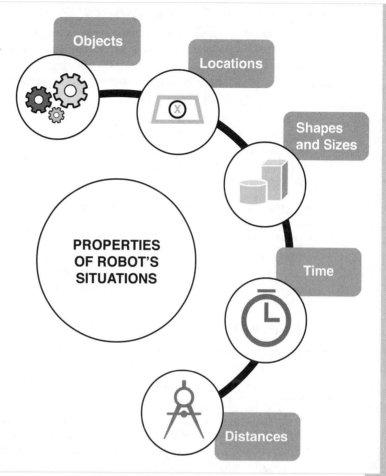

Level 6 Task Vocabulary

A level 6 vocabulary is similar to a level 4 vocabulary in that it describes actions and capabilities of the robot. The difference is that level 6 vocabularies describe actions and capabilities that are situation specific.

For example, if Midamba had access to our Unit2 robot with the hardware listed in the capability matrix of Table 2.1, an appropriate task vocabulary would support instructions like

- Action 1. Scan the shelves for a blue can of oil.

- Action 2. Measure the grade of the oil.

- Action 3. Retrieve the can of oil if it is grade A and contains at least 2 quarts.

 note

Notice all the actions involve vocabulary specific to a particular situation. So the level 6 task vocabulary can be described as a *situation specific* level 4 vocabulary.

Level 7 Scenario Vocabulary

Our novice robot programmer Midamba has a scenario where one or more robots in the research area must identify some kind of chemical that can help him charge his primary battery or neutralize the acid on his spare battery. Once the chemical is identified, the robots must deliver it to him. The sequence of tasks in the *situations* make up the *scenario*. The scenario vocabulary for Unit2 would support a scenario description like the following:

> Unit2, from your current position scan the shelves
> until you find a chemical that can either charge
> my primary battery or clean the acid from my
> spare battery and then retrieve that chemical
> and return it to me.

Ideally, the level 7 robot vocabulary would allow you to program the robot to accomplish this set of tasks in a straightforward manner. Utilitarian autonomous robots have to have some kind of level 7 vocabulary to be useful and effective. The relationship between robot vocabulary levels 5 through 7 is shown in Figure 2.7.

The vocabulary of levels 5 through 7 allows the programmer to describe the robot's role in a given situation and scenario and how the robot is to accomplish that role.

What's Ahead?

The first part of this book focuses on programming robot capabilities, such as sensor, movement, and end-effector programming. The second part of this book focuses on programming robots to execute roles in particular situations or scenarios. Keep in mind you can program a robot at any of these levels.

In some cases, some programming may already be done for you. But truly programming an autonomous robot to be successful in some scenario requires an appropriate ROLL model and implementation. Before we describe in more detail Midamba's robots and their scenarios, we describe how to RSVP your robot in Chapter 3, "RSVP: Robot Scenario Visual Planning."

Figure 2.7
The relationships between levels 5 through 7

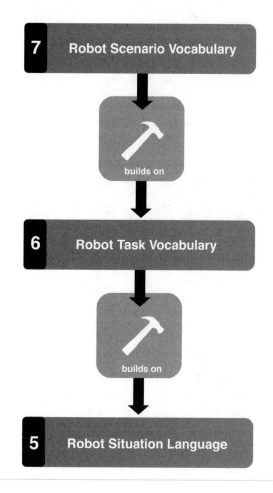

3

RSVP: ROBOT SCENARIO VISUAL PLANNING

Robot Sensitivity Training Lesson #3: *Don't instruct the robot to perform a task you can't picture it performing.*

As described in Chapter 2, "Robot Vocabularies," the *robot vocabulary* is the language you use to assign a robot tasks for a specific situation or scenario. And once a vocabulary has been established, figuring out the instructions for the robot to execute using that vocabulary is the next step.

Making a picture or a "visual representation" of the scenario and instructions you want the robot to perform can be great way to ensure your robot performs the tasks properly. A picture of the instructions the robot will perform allows you to think through the steps before translating them to the code. Visuals can help you understand the process, and studying that visual can improve its development by seeing what has to be done and elucidating that which may otherwise pose a problem. We call this the RSVP (Robot Scenario Visual Planning). The RSVP is a visual that helps develop the plan of instructions for what the robot will do. The RSVP is composed of three types of visuals:

- A floorplan of the physical environment of the scenario

- A statechart of the robot and object's states

- Flowcharts of the instructions for the tasks

These visuals ensure that you have a "clear picture" of what has to be done to program a robot to do great feats that can save the world or light the candles on a cake. RSVP can be used in any combination. Flowcharts may be more useful than statecharts for some. For others, statecharts are best. All we suggest is that a floorplan or layout is needed whether statecharts or flowcharts are utilized.

The saying "a picture is worth a thousand words" means that a single image can convey the meaning of a complex idea as well as a large amount of descriptive text. We grew up with this notion while in grade school especially when trying to solve word problems; "draw a picture" of the main ideas of the word problem and magically it becomes clear how to solve it. That notion still works. In this case, drawing a picture of the environment, a statechart, and flowcharts will be worth not only a thousand words but a thousand commands. Developing an RSVP allows you to plan your robot navigation through your scenario and work out the steps of the instructions for the tasks in the various situations. This avoids the trials and errors of directly writing code.

Mapping the Scenario

The first part of the RSVP is a map of the scenario. A map is a symbolic representation of the environment where the tasks and situations will take place. The environment for the scenario is the world in which the robots operate. Figure 3.1 shows the classic Test Pad for NXT Mindstorms robot.

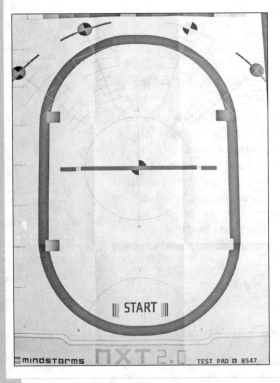

Figure 3.1
A robot world for NXT Mindstorms Test Pad

A Test Pad like the one shown in Figure 3.1 is part of the Mindstorms robot kits. This Test Pad is approximately 24 inches wide, 30 inches long, and has a rectangular shape. There are 16 colors on the Test Pad and 38 unique numbers with some duplicates. There is a series of straight lines and arcs on the pad. Yellow, blue, red, and green squares are on the Test Pad along with other colored shapes in various areas on the pad. It is the robot's world or environment used for the initial testing of NXT Mindstorms robots' color sensors, motors, and so on.

Like the Test Pad, a floorplan shows the locations of objects that are to be recognized like colored squares, objects the robot will interact with, or obstacles to be avoided. If objects are too high or too far away, sensors may not be able to determine their location. Determining the path the robot must navigate to reach those locations can also be planned by using this map.

The dimension of the space and of the robot (the robot footprint) may affect the capability of the robot to navigate the space and perform its tasks. For example, for our BR-1 robot, what is the location of the cake relative to the location of the robot? Is there a path? Are there obstacles? Can the robot move around the space? This is what the map helps determine.

 tip

Next to the actual robot, the robot's environment is the most important consideration.

Creating a Floorplan

The map can be a simple 2D layout or floorplan of the environment using geometric shapes, icons, or colors to represent objects or robots. For a simple map of this kind, depicting an accurate scale is not that important, but objects and spaces should have some type of relative scale.

Use straight lines to delineate the area. Decide the measurement system. Be sure the measurement system is consistent with the API functions. Use arrows and the measurements to mark the dimensions of the area, objects, and robot footprint. It's best to use a vector graphics editor to create the map. For our maps we use Libre Office Draw. Figure 3.2 shows a simple layout of a floorplan of the robot environment for BR-1.

In Figure 3.2, the objects of interest are designated: locations of the robot, the table, and the cake on the table. The floorplan marks the dimensions of the area and the footprint of the robot. The lower-left corner is marked (0,0) and the upper-right corner is marked (300,400). This shows the dimensions of the area in cm. It also marks distances between objects and BR-1. Although this floorplan is not to scale, lengths and widths have a relative relationship. BR-1's footprint length is 50 cm and width is 30 cm.

BR-1 is to light the candles on the cake. The cake is located at the center of an area that is 400 cm × 300 cm. The cake has a diameter of 30 cm on a table that is 100 cm × 100 cm. That means the robot arm of BR-1 should have a reach of at least 53 cm from the edge of the table to reach the candle at the farthest point in the X dimension.

The maximum extension of the robot arm to the tip of the end-effector is 80 cm, and the length of the lighter adds an additional 10 cm. The task also depends on some additional considerations:

- The height of the candle

- The height of the cake

- The length of BR-1 from the arm point to the top of the candle wick

- The location of the robot

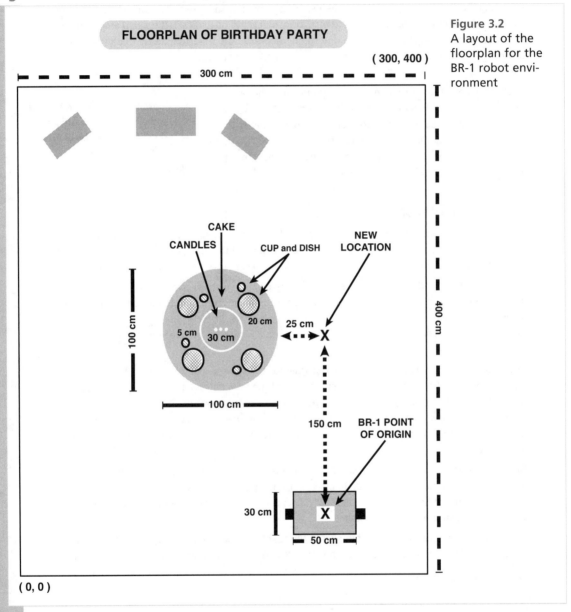

FLOORPLAN OF BIRTHDAY PARTY

(300, 400)

300 cm

CAKE

CANDLES

CUP and DISH

NEW LOCATION

100 cm

20 cm

25 cm

5 cm

30 cm

X

100 cm

400 cm

150 cm

BR-1 POINT OF ORIGIN

30 cm

X

50 cm

(0, 0)

Figure 3.2

A layout of the floorplan for the BR-1 robot environment

Figure 3.3 shows how to calculate the required reach to light the candle. In this case, it is the hypotenuse of a right triangle. Leg "a" of the triangle is the height of the robot from the top of the wick to the robot arm joint which is 76 cm, and leg "b" is the radius of the table plus the 3 cm to the location of the farthest candle on the cake, which is 53 cm.

So the required reach of the robot arm, end-effector, and lighter is around 93 cm. But the robot's reach is only 90 cm. So BR-1 will have to lean a little toward the cake or get a lighter that is 3 cm longer to light the wick.

note

Determining the positions and required extension of a robot arm is far more complicated than this simple example and is discussed in Chapter 9, "Robot SPACES." But what is important in the example is how the layout/floorplan helps elucidate some important issues so that you can plan your robot's tasks.

Figure 3.3
Calculating the length of the robot arm as the hypotenuse of a right triangle

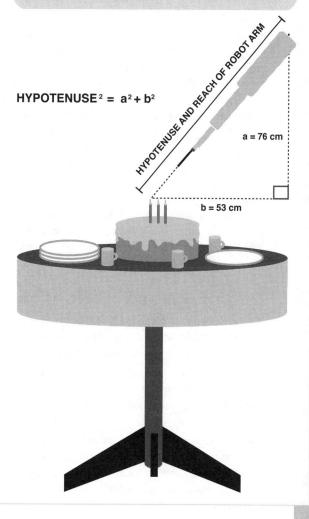

CALCULATING LENGTH OF ROBOT ARM

$$\text{HYPOTENUSE}^2 = a^2 + b^2$$

HYPOTENUSE AND REACH OF ROBOT ARM

$a = 76$ cm

$b = 53$ cm

The Robot's World

For the robot to be automated it requires details about its environment. Consider this: If you are traveling to a new city you know nothing about, how well will you be able to do the things you want to do? You do not know where anything is. You need a map or someone to show you around and tell you "here is a restaurant" and "here is a museum." A robot that is *fully automated* must have sufficient information about the environment. The more information the robot has, the more likely the robot can accomplish its goal.

 note

The robot's world is the environment where the robot performs it tasks. It's the only world the robot is aware of. Nothing outside that environment matters, and the robot is not aware of it.

All environments are not alike. We know environments are dynamic. The robot's environments can be partially or fully accessible to a robot. A fully accessible environment means all objects and aspects of the environment are within the reach of the robot's sensors. No object is too high, low, or far away from the robot to detect or interact with. The robot has all the necessary sensors to receive input from the environment. If there is a sound, the robot can detect it with its sound sensor. If a light is on, the robot can detect it with its light sensor.

A partially accessible environment means there are aspects of the environment the robot cannot detect or there are objects the robot cannot detect or interact with because it lacks the end-effector to pick it up or the location sensor to detect it. An object that is 180 cm from the ground is out of the reach of the robot with a 80 cm arm extension and a height of 50 cm. What if BR-1 is to light the candles once the singing begins and it does not have a sound sensor? Sound is part of the environment; therefore, it will not be able to perform the task. So when creating the floorplan for a partially accessible environment, consider the "robot's perspective." For example, for objects that are not accessible by the robot, use some visual indicator to distinguish those for the objects the robot can access. Use color or even draw a broken line around it.

Deterministic and Nondeterministic Environments

What about control? Does the robot control every aspect of its environment? Is the robot the only force that controls or manipulates the objects in its environment? This is the difference between a *deterministic* and *nondeterministic* environment.

With a deterministic environment, the next state is completely determined by the current state and the actions performed by the robot(s). This means if the BR-1 robot lights the candles, they will stay lit until BR-1 blows them out. If BR-1 removes the dishes from the table, they will stay in the location they're placed.

With a nondeterministic environment, like the one for the birthday party scenario, BR-1 does not blow out the candles. (It would be pretty mean if it did.) Dishes can be moved around by the attendees of the party, not just BR-1. What if there are no obstacles between BR-1 and its destination and then a partygoer places an obstacle there? How can BR-1 perform its tasks in a dynamic nondeterministic environment?

Each environment type has its own set of challenges. With a dynamic nondeterministic environment, the robot is required to consider the previous state and the current state before a task is attempted and then make a decision whether the task can be performed.

Table 3.1 lists some of the types of environments with a brief description.

Table 3.1 Some Types of Environments with a Brief Description

Environment Type	Description
Fully accessible	All aspects of the environment are accessible through the robot's sensors, actuators, and end-effectors.
Partially accessible	Some objects are not accessible or cannot be sensed by the robot.
Deterministic	The next state of the environment is completely determined by the current state and actions performed by the robot.
Nondeterministic	The next state of the environment is not completely under the control of the robot; the object may be influenced by outside factors or external agents.

RSVP READ SET

Many aspects of the environment are not part of the layout or floorplan but should be recorded somehow to be referenced when developing the instructions for the tasks. For example, the color, weight, height, and even surface type of the objects are all detectable characteristics that are identified by sensors or affect motors and end-effectors as well as the environment type, identified outside forces, and their impact on objects.

Some of these characteristics can be represented in the floorplan. But a READ set can contain all the characteristics. Each type of environment should have its own READ set.

For example, color is a detectable characteristic identified by a color or light sensor. The object's weight determines whether the robot can lift, hold, or carry the object to another location based on the torque of the servos. The shape, height, and even the surface determine whether the object can be manipulated by the end-effector.

 note

A READ (Robot Environmental Attribute Description) set is a construct that contains a list of objects that the robot will encounter, control, and interact with within the robot's environment. It also contains characteristics and attributes of the objects detectable by the robot's sensors or that affect how the robot will interact with that object.

Any characteristic of the environment is part of the READ set, such as dimensions, lighting, and terrain. These characteristics can affect how well sensors and motors work. The lighting of the environment, whether sunlight, ambient room light, or candle light, affects the color and light sensor differently. A robot traveling across a wooden floor is different from the robot traveling across gravel, dirt, or carpet. Surfaces affect wheel rotation and distance calculations.

Table 3.2 is the READ set for the Mindstorms NXT Test Pad.

Table 3.2 READ Set for the Mindstorms NXT Test Pad

Object: Physical Work Space

Attribute	Value
Environment type	Deterministic, fully accessible
Width	24 inches
Length	30 inches
Height	0
Shape	Rectangular
Surface	Paper (smooth)

Object: Color (Light)

Attribute	Value
Num of colors	16
Light intensities	16
Colors	Red, green, blue, yellow, orange, white, black, gray, green, light blue, silver, etc.

Object: Symbols

Attribute	Value
Symbol	Integers
Integer values	0–30, 90, 120, 180, 270, 360, 40, 60, 70
Geometric	Lines, arcs, squares

The READ set for the Test Pad describes the workspace including its type (fully accessible and deterministic), all the colors, and symbols. It describes what will be encountered by a robot when performing a search, such as identifying the blue square. The sets list the attributes and values of the physical workspace, colors, and symbols on the Test Pad.

For a dynamic environment such as our birthday party scenario, the READ set can contain information pertaining to the outside forces that might interact with the objects. For example, there are initial locations for the dishes and cups on the table, but the partygoers may move their dishes and cups to a new location on the table. The new locations should be represented in the READ set along with the time or the condition this occurred. Once the party is over and BR-1 is to remove those dishes and cup, each location should be updated. Table 3.3 is the READ set for the birthday party for the BR-1.

Table 3.3   READ Set for the Birthday Party Scenario

Object: Physical Work Space

Attribute	Value	Force	Time/Condition	New Value
Environment type	Nondeterministic partially			
Width	300 cm			
Length	400 cm			
Height	0			
Shape	Rectangular			
Surface	Paper (smooth)			
Lighting	Artificial			

Object: Cake

Attribute	Value	Force	Time/Condition	New Value
Height	14 cm			
Diameter	30 cm			
Location	150, 200	External	N/A	
Placement	Table	External	N/A	
Related objects	Candles			

Object: Candles

Attribute	Value	Force	Time/Condition	New Value
Height	4 cm			
Number of candles	3			
Locations	1 153, 200 2 150, 200 3 147, 200	External	N/A	
Condition 1	Unlit	BR-1	Singing starts	Lit
Condition 2	Lit	External	Singing ends	Unlit

Object: Dishes

Attribute	Value	Force	Time/Condition	New Value
Diameter	20 cm			
Height	1 cm			
Number of dishes	4			
Locations	1 110, 215 2 110, 180 3 170, 215 4 170, 180	External	After party ends	All at 110, 215 (stacked) Height 2 cm

Object: Cups				
Attribute	Value	Force	Time/Condition	New Value
Diameter	5 cm			
Height	10 cm			
Number of dishes	4			
Locations	1 119, 218	External	After party ends	All at 119, 218 (stacked)
	2 105, 189			Height 14 cm
	3 165, 224			
	4 163, 185			

This READ set has three additional columns:

- Force

- Time/Condition

- New Value

Force is the source of the iteration with the object; this force is anything working in the environment that is not the robot. The *Time/Condition* denotes when or under what condition the force interacts with the object. The *New Value* is self-explanatory.

Pseudocode and Flowcharting RSVP

Flowcharting is an RSVP used to work out the flow of control of an object to the whole system. It is a linear sequence of lines of instructions that can include any kind of looping, selection, or decision-making. A flowchart explains the process by using special box symbols that represent a certain type of work. Text displayed within the boxes describes a task, process, or instruction.

Flowcharts are a type of statechart (discussed later in this chapter) since they also contain states that are converted to actions and activities. Things like decisions and repetitions are easily represented, and what happens as the result of a branch can be simply depicted. Some suggest flowcharting before writing pseudocode. Pseudocode has the advantage of being easily converted to a programming language or utilized for documenting a program. It can also be easily changed. A flowchart requires a bit more work to change when using flowcharting software.

Table 3.4 list advantages and disadvantages of pseudocode and flowcharting. Both are great tools for working out the steps. It is a matter of personal taste which you will use at a particular time in a project.

Table 3.4   Advantages and Disadvantages of Pseudocode and Flowcharting

RSVP Type	Advantages	Disadvantages
Pseudocode: A method of describing computer instructions using a combination of natural language or programming language.	Easily created and modified in any word processor. Implementation is useful in any design. Written and understood easily. Easily converted to a programming language.	Is not visual. No standardized style or format. More difficult to follow the logic.
Flowcharting: Flow from the top to the bottom of a page. Each command is placed in a box of the appropriate shape, and arrows are used to direct program flow.	Is visual, easier to communicate to others. Problems can be analyzed more effectively.	Can become complex and clumsy for complicated logic. Alterations may require redrawing completely.

The four common symbols used in flowcharting are

- **Start and stop:** The start symbol represents the beginning of the flowchart with the label "start" appearing inside the symbol. The stop symbol represents the end of the flowchart with the label "stop" appearing inside the symbol. These are the only symbols with keyword labels.

- **Input and output:** The input and output symbol contains data that is used for input (e.g., provided by the user) and data that is the result of processing (output).

- **Decisions:** The decision symbol contains a question or a decision that has to be made.

- **Process:** The process symbol contains brief descriptions (a few words) of a rule or some action taking place.

Figure 3.4 shows the common symbols of flowcharting.

Each symbol has an inbound or outbound arrow leading to or from another symbol. The start symbol has only one outbound arrow, and the stop symbol has only one inbound arrow. The "start" symbol represents the beginning of the flowchart with the label "start" appearing inside the symbol.

The "stop" symbol represents the end of the flowchart with the label "stop" appearing inside the symbol. These are the only symbols with keyword labels. The decision symbol will contain a question or a decision that has to be made. The process symbol will contain brief descriptions (a few words) of a rule or some action taking place. The decision symbol has one inbound arrow and two outbound arrows. Each arrow represents a decision path through the process starting from that symbol:

- TRUE/YES

- FALSE/NO

COMMON FLOWCHART SYMBOLS

Figure 3.4
The common symbols of flowcharting

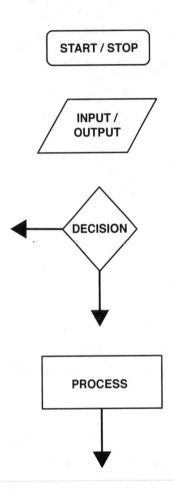

The process, input, and output symbols have one inbound and one outbound arrow. The symbols contain text that describes the rule or action, input or output. Figure 3.5 shows the "Lighting candles" flowchart.

Notice at the beginning of the flowchart, below the start symbol, BR-1 is to wait until the singing begins. A decision is made on whether the singing has started. There are two paths: If the singing has not started, there is a FALSE/NO answer to the question and BR-1 continues to wait. If the singing has started, there is a TRUE/YES answer and BR-1 enters a loop or decision.

Figure 3.5
The lighting candles flowchart

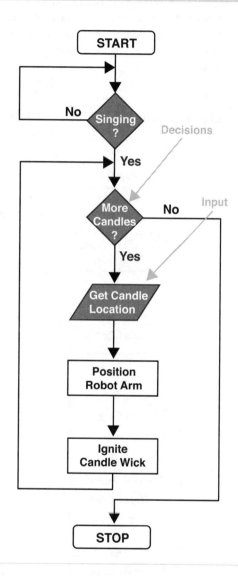

If there are candles to light, that is the decision. If yes, it gets the position of the next candle, positions the robot arm to the appropriate position to ignite the wick, and then ignites the wick. An input symbol is used to receive the position of the next candle to light. The BR-1 is to light all the candles and stops once complete.

Flow of Control and Control Structures

The task a robot executes can be a series of steps performed one after another, a sequential flow of control. The term *flow of control* details the direction the process takes, which way program control "flows." Flow of control determines how a computer responds when given certain conditions and parameters. An example of sequential flow of control is in Figure 3.6. Another robot in our birthday scenario is BR-3. Its task is to open the door for the guests. Figure 3.6 shows the sequential flow of control for this task.

BR-3's SEQUENTIAL FLOWCHART

Figure 3.6
The flowchart for BR-3 ·

The robot goes to the door, opens it, says, "Welcome," and then closes the door and returns to its original location. This would look like a rather inconsiderate host. Did the doorbell ring, signaling BR-3 that guests were at the door? If someone was at the door, after saying "Welcome," did BR-3 allow the guest to enter before closing the door? BR-3 should be able to act in a predictable way at the birthday party. That means making decisions based on events and doing things in repetition.

A decision symbol is used to construct branching for alternative flow controls. Decision symbols can be used to express decision, repetition, and case statements. A simple decision is structured as an if-then or if-then-else statement.

A simple if-then decision for BR-3 is shown in Figure 3.7 (a). "If Doorbell rings, then travel to door and open it." Now BR-3 will wait until the guest(s) enters before it says "Welcome." Notice the alternative action to be taken if the guest(s) has not entered. BR-3 will wait 5 seconds and then check if the guest(s) has entered yet. If Yes then BR-3 says "Welcome" and closes the door. This is shown in Figure 3.7 (b) if-then-else; the alternative action is to wait.

Figure 3.7
The flowchart for if-then and if-then-else decisions

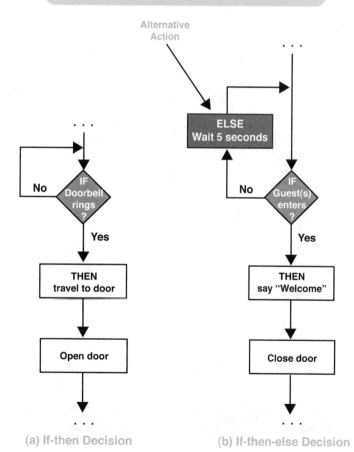

In Figure 3.7, the question (or condition test) to be answered is whether the doorbell has rung. What if there is more than one question/condition test that has to be met before BR-3 is to open the door? With about BR-1, what if there were multiple conditions that had to be met before lighting the candles:

- "If there is a singing AND the lighter is lit then light the candles."

- In this case, both conditions have to be met. This is called a *Nested* decision or condition.

What if there is a question or condition in which there are many different possible answers and each answer or condition has a different action to take? For example, what if as our BR-1 or BR-3 travels across the room it encounters an object and has to maneuver around the object to reach its destination. It could check the range of the object in its path to determine the action to take to avoid it. If the object is within a certain range, BR-1 and BR-3 turn to the left either 90 degrees or 45 degrees, travel a path around the object, and then continue on their original path to their destinations as shown in Figure 3.8.

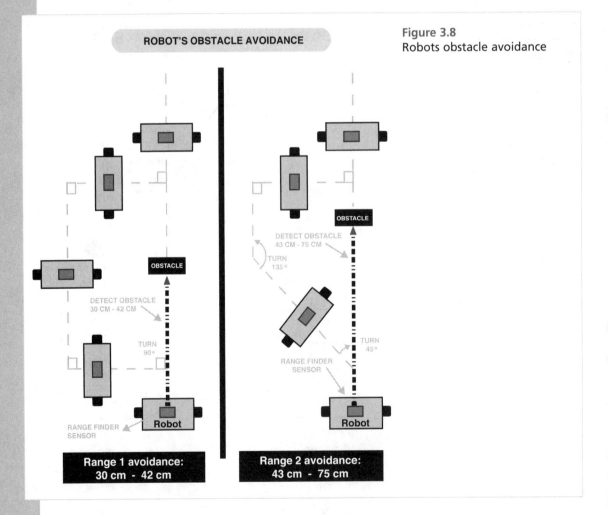

Figure 3.8
Robots obstacle avoidance

Using the flowchart, this can be expressed as a series of decisions or a case statement. A case is a type of decision where there are several possible answers to a question. With the series of decisions, the same question is asked three times, each with a different answer and action. With a case statement, the question is expressed only one time. Figure 3.9 contrasts the series of decisions in the case statement, which is simpler to read and understand what is going on.

Figure 3.9
Contrast case statement
from a series of decisions

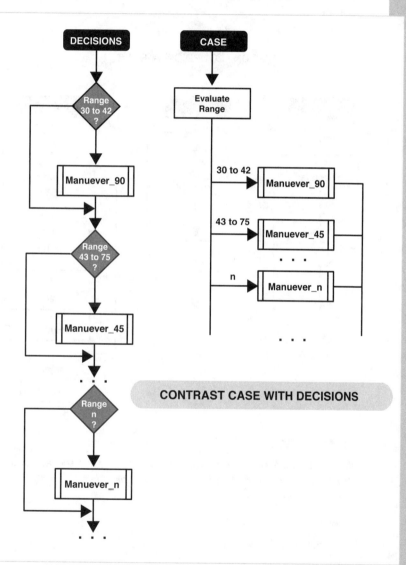

Repetition or looping is shown in Figure 3.10. In a loop, a simple decision is coupled with an action that is performed before or after the condition test. Depending on the result, the action is performed again. In Figure 3.10 (a), the action will be performed at least once. If the condition is not met (singing has not started—maybe everyone is having too much fun), the robot must continue to wait. This is an example of a do-until loop, "do" this action "until" this condition is true. A while loop performs the condition test first and if met, then the action is performed. This is depicted in Figure 3.10 (b), while singing has not started, wait. BR-1 will loop and wait until singing starts, as in the do-until loop. The difference is a wait is performed after the condition is met. Another type is the for loop, shown in Figure 3.10 (c), where the condition test controls the specific number of times the loop is executed.

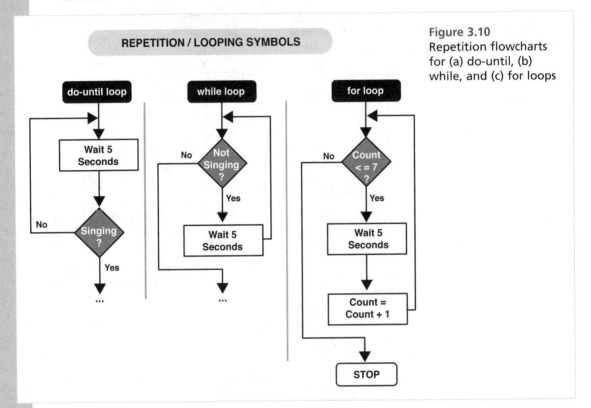

REPETITION / LOOPING SYMBOLS

Figure 3.10
Repetition flowcharts
for (a) do-until, (b)
while, and (c) for loops

Subroutines

When thinking about what role your robot is to play in a scenario or situation, the role is broken down into a series of actions. BR-1's role is to be a host at a birthday party. This role is broken down into four states:

- Idle
- Traveling
- Lighting candles
- Waiting
- Removing dishes

This can be broken down into a series of actions or tasks:

1. Wait until singing begins.

   - Travel to birthday cake table.

   - Light the candles on the cake.

   - Travel to the original location.

2. Wait until party is over.

   - Remove dishes from cake table.

   - Travel back to original location.

These are short descriptions of tasks. Each task can be further broken down into a series of steps or subroutines. "Lighting candles" is a composite state that is broken down into other substates:

- Locating wick

- Igniting wick

Actually, "Remove dishes from cake table" and "Travel back to original location" should also be broken down into subroutines. Removing dishes from the cake table requires the positioning of the robot arm to remove each plate and cup subroutines, and traveling requires the rotating of motors subroutines.

Figure 3.11 shows the flowcharting for LightingCandles and its subroutines LocatingWick and IgnitingWick.

A subroutine symbol is the same as a process symbol, but it contains the name of the subroutine with a vertical line on each side of the name of the subroutine. The name of a subroutine can be a phrase that describes the purpose of the subroutine.

Flowcharts are then developed for those subroutines. What's great about using subroutines is the details don't have to be figured out immediately. Figuring out how the robot will perform a task can be put off for a while. The highest level processes can be worked out and then later actions/tasks can be broken down.

A subroutine can be identified and generalized from similar steps used at different place, in the robot's process. Instead of repeating a series of steps or developing different subroutines, the process can be generalized and placed in one subroutine that is called when needed. For example, the traveling procedure started out as a series of steps for BR-1 to travel to the cake table (TableTravel) and then a series of steps to travel back to its original location (OriginTravel). These are the same tasks with different starting and ending locations. Instead of subroutines that use the starting and ending locations, a Travel subroutine requires both the current and final locations of the robot to be used.

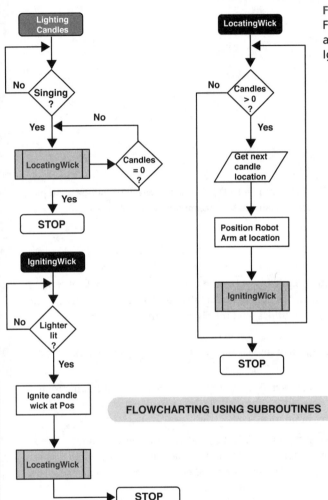

Figure 3.11
Flowcharting for LightingCandles and its subroutines LocatingWick and IgnitingWick

FLOWCHARTING USING SUBROUTINES

Statecharts for Robots and Objects

A statechart is one way to visualize a state machine.

For example, a "change of state" can be as simple as a change of location. When the robot travels from its initial location to the location next to the table, this is a change of the robot's state. Another example is that the birthday candles change from an unlit state to a lit state. The state machine captures the events, transformations, and responses. A statechart is a diagram of these activities. The statechart is used to capture the possible

 note

A state machine models the behavior of a single robot or object in an environment. The states are the transformations the robot or object goes through when something happens.

situations for that object in that scenario. As you recall from Chapter 2, a situation is a snapshot of an event in the scenario. Possible situations for the BR-1 are

- **Situation 1:** BR-1 waiting for signal to move to new location

- **Situation 2:** BR-1 traveling to cake table

- **Situation 3:** BR-1 next to cake on a table with candles that have not yet been lit

- **Situation 4:** BR-1 positioning the lighter over the candles, and so on

All these situations represent changes in the state of the robot. Changes in the state of the robot or object take place when something happens, an event. That event can be a signal, the result of an operation, or just the passing of time. When an event happens, some activity happens, depending on the current state of the object. The current state determines what is possible.

The event works as a trigger or stimulus causing a condition in which a change of state can occur. This change from one state to another is called a transition. The object is transitioning from stateA, the source state, to stateB, the target state. Figure 3.12 shows a simple state machine for BR-1.

Figure 3.12
State machine for BR-1

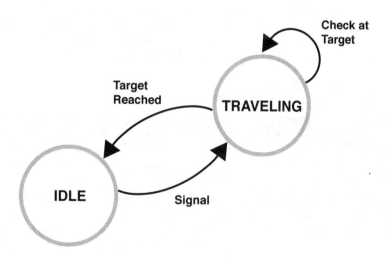

SIMPLE BR-1 STATE MACHINE

Check at Target

Target Reached

TRAVELING

IDLE   Signal

Figure 3.12 shows two states for BR-1: Idle or Traveling. When BR-1 is in an Idle state, it is waiting for an event to occur. This event is a signal that contains the new location for the robot. Once the robot receives this signal, it transitions from Idle to Traveling state. BR-1 continues to travel until it reaches its target location. Once reached, the robot transitions from the Traveling state back to an Idle state. Signals, actions, and activities may be performed or controlled by the object or by outside forces. For example, the new location will not be generated by BR-1 but by another agent. BR-1 does have the capability to check its location while traveling.

Developing a Statechart

As discussed earlier, a state is condition or situation of an object that represents a transformation during the life of the object. A state machine shows states and transitions between states. There are many ways to represent a state machine. In this book, we represent a state machine as a UML (Unified Modeling Language) statechart. Statecharts have additional notations for events, actions, conditions, parts of transitions, and types and parts of states.

There are three types of states:

 note

In a statechart, the nodes are states and the arcs are transitions. The states are represented as circles or as rounded-corner rectangles in which the name of the state is displayed. The transitions are arcs that connect the source and the target state with an arrow pointing to the target state.

- **Initial:** The default starting point for the state machine. It is a solid black dot with a transition to the first state of the machine.

- **Final:** The ending state, meaning the object has reached the end of its lifetime. It is represented as a solid dot inside a circle.

- **Composite state and substate:** A state contained inside another state. That state is called a superstate or composite state.

States have different parts. Table 3.5 lists the parts of states with a brief description. A state node displaying its name can also display the parts listed in this table. These parts can be used to represent processing that occurs when the object transitions to the new state. There may be actions to take as soon as the object enters and leaves the state. There may be actions that have to be taken while the object is in a particular state. All this can be noted in the statechart.

Table 3.5 Parts of a State

Part	Description
Name	The unique name of the state distinguishes it from other states.
Entry/exit actions	Actions executed when entering the state (entry actions) or executed when exiting the state (exit actions).
Composite/substates	A nested state; the substates are the states that are activated inside the composite state.
Internal transitions	Transitions that occur within the state are handled without causing a change in the state but do not cause the entry and exit actions to execute.
Self-transitions	Transitions that occur within the state that are handled without causing a change in the state. They cause the exit and entry actions to execute when exiting and then reentering the state.

Figure 3.13 shows a state node and format for actions, activities, and internal transition statements.

Figure 3.13
A state node and the format
of statements

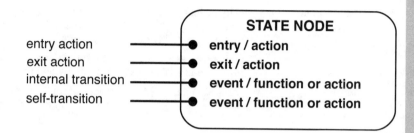

STATE NODE STATEMENT FORMATS

entry action ——————
exit action ——————
internal transition ——————
self-transition ——————

STATE NODE
● **entry / action**
● **exit / action**
● **event / function or action**
● **event / function or action**

The entry and exit action statements have this format:

- Entry/action or activity

- Exit/action or activity

This is an example of an entry and exit action statement for a state called Validating:

- entry action: entry / validate(data)

- exit action: exit   / send(data)

Upon entering the Validating state, the validate(data) function is called. Upon exiting this state, the exit action send(data) is called.

Internal transitions occur inside the state. They are events that take place after entry actions and before exit actions if there are any. Self-transitions are different from internal transitions. With a self-transition, the entry and exit actions are performed. The state is left; the exit action is performed.

Then the same state is reentered and the entry action is performed. The action of the self-transition is performed after the exit action and before the entry action. Self-transitions are represented as a directed line that loops and points back to the same state.

An internal or self-transition statement has this format:

- Name/action or function

For example:

- do / createChart(data)

"do" is the label for the activity, the function "createChart(data)" is executed.

There are several parts of a transition, the relationship between two states. We know that triggers cause transitions to occur, and actions can be coupled with triggers. A met condition can also cause a transition. Table 3.6 lists the parts of a transition.

Table 3.6   Parts of a Transition

Part	Description
Source state	The original state of the object; when a transition occurs the object leaves the source state.
Target state	The state the object enters after the transition.
Event trigger	The event that causes the transition to occur. A transition may be triggerless, which means the transition occurs as soon as the object completes all activities in the source state.
Guard condition	A boolean expression associated with an event trigger, which when evaluated to TRUE, causes the transition to take place.
Action	An action executed by the object that takes place during a transition; may be associated with an event trigger or guard condition.

An event trigger has a similar format as a state action statement:

- Name/action or function

- name [Guard] / action or function

For example, for the internal transition statement, a guarded condition can be added:

- do [Validated] / createChart(data)

Figure 3.14 is the statechart for BR-1.

There are four states: Idle, Traveling, Lighting candles, Waiting and Removing dishes. When transitioning from Idle to Traveling, BR-1 gets the new location and knows its mission:

- do [GetPosition] / setMission()

There are two transitions from the Traveling state:

- Traveling to Lighting candles

- Traveling to Removing dishes

 note

A guard condition is a boolean value or expression that evaluates to True or False. It is enclosed in square brackets. The guard condition has to be met for the function to execute. It can be used in a state or transition statement.

 note

Validated is a boolean value. It is a condition that has to be met for createChart() to execute.

Traveling transitions to LightingCandles when its target is reached and its Mission is *candles*. Traveling transitions to Removing dishes when its target is reached and its Mission is *dishes*. To transition from LightingCandles, "candles" mission must be complete. To transition from RemovingDishes to the final state, all missions must be completed.

LightingCandles is a composite state that contains two substates: LocatingWick and IgnitingWick. Upon entering the Lighting candles state, the boolean value Singing is evaluated. If there is singing, then the candles are to be lit. First the wick has to be located, then the arm is moved to that location, and finally the wick can be lit. In the LocatingWick state, the entry action evaluates the expression:

- Candles > 0

Figure 3.14
Statechart for BR-1

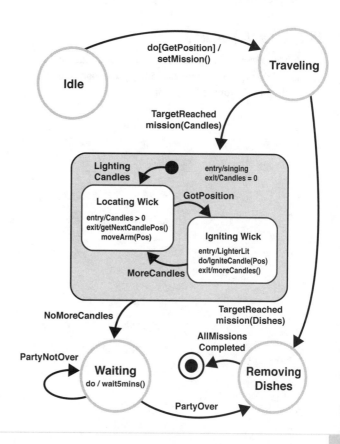

STATECHART FOR BR-1'S TASKS

If True, the state exits when the position of the first or next candle is retrieved and then the robot arm is moved to the position (Pos).

The position of the wick is retrieved, so BR-1 transitions to "IgnitingWick." Upon entry, the lighter is checked to see if it is lit. If lit the candle wick is lit, (an internal state). To exit this state, Candles > 0, then "LocatingWick" state is reentered. If Candles = 0, then BR-1 transitions to "Waiting" state. BR-1 waits until the party is over. Then BR-1 can remove all the dishes. In the "Waiting" state, there is a self-transition "PartyNotOver." Remember, with a self-transition, the exit and entry actions are performed as the state is exited and reentered. In this case, there are no exit actions, but there is an entry action "wait 5 minutes". The guard condition is checked, "PartyNotOver." If the party is not over, then the state is reentered and the entry action is executed; BR-1 waits for 5 minutes. Once the party is over, then BR-1 transitions to "RemovingDishes." This is the last state. If boolean value

AllMissionsCompleted is True, BR-1 transitions to the final state. But some objects may not have a final state and work continuously. Statecharts are good for working out the changing aspect of an object during the lifetime of the object. It shows the flow of control from one state of the object to another state.

What's Ahead?

In Chapter 4, "Checking the Actual Capabilities of Your Robot," we discuss what your robot is capable of doing. This means checking the capabilities and limits of the microcontroller, sensors, motors, and end-effectors of the robots.

CHECKING THE ACTUAL CAPABILITIES OF YOUR ROBOT

Robot Sensitivity Training Lesson #4: *Some robots underperform only because their potential has been overpromised.*

Understanding the robot's real capabilities is a prerequisite to coming up with the robot's vocabularies, ROLL model, or any useful programming. The robot capability matrix is designed to keep track of what parts and capabilities a robot has.

Our first cut at a robot capability matrix from Table 2.1 listed the sensors, end-effectors, microcontroller, and type of robot, but that only tells part of the story. Some of the capabilities were taken directly from the manufacturers' specifications. But a note of caution is in order. The specifications for the robot's sensors, actuators, motors, and controllers are often one thing on paper and something different during robot use. Take, for example, a sensor's specification. There is a certain amount of error or limit to the precision of any sensor or actuator; for example:

- Actual measurable distance by a sensor is less than specified measurable difference.

- A light sensor is specified to recognize three colors but actually only recognizes two colors.

The error rate for a sensor may not be stated clearly or may be totally missing in the documentation. Not all motors are created equal, and there is a great deal of variance in the level of precision an actuator may support.

Since sensors, actuators, and end-effectors can be controlled by programming, performance levels and limitations have to be identified prior to the programming effort. Manufacturers are sometimes overly optimistic about the performance of a controller, sensor, or actuator, and the actual capability turns out to be different from the specified capability. In other instances, there are simply mistakes or typos in the documentation; for example, the documentation specifies kilograms for actuator capability when grams should have been used.

 note

Remember, the actuators/ motors ultimately determine how fast a robot can move. A robot's acceleration is tied to its actuators. The actuators are also responsible for how much a robot can lift or how much a robot can hold.

It is a waste of effort and could result in damage to the robot if instructions given to the robot rely on manufacturer specifications that were not correct. For example, actuators and motors provide most of the work required for robot arms. So instructing a robot arm to lift and relocate 2 liters when its actuators can only handle 1.8 liters could result in damage to the robot's servos or present some kind of safety issue.

Engaging robot actuators with too much weight is a common cause of robot failure. Over specified sensors are another common cause of robot failure. Relying on a robot's sensors in situations that call for the identification of objects of different sizes, shapes, and colors when the scanning sensor can only recognize one shape and one color is typical of the kinds of mismatches found between sensor specifications and actual sensor performance.

Before the real programming effort begins we have to check or test the components of the basic robot skeleton shown in Figure 4.1. These components are the hardware foundation of robot programming. Any one of these components can be a show-stopper, so it's important to know the basic limits of each. The microcontroller contains the microprocessor, input/output ports, sensor ports, and various types of memory, flash RAM, and EEPROM. Figure 4.2 contains the basic layout of a microcontroller.

 note

An EEPROM, or electrically erasable programmable read-only memory, like a regular ROM chip, uses a grid and electrical impulses to create binary data. However, the difference between ROM chips and EEPROM chips is that EEPROM chips can be reprogrammed without removing them from the computer, contrary to basic ROM chips, which can only be programmed one time.

 tip

A robot may have multiple controllers. It's important to know the actual specifications. A robot may have one controller for motors and another controller for servos and yet another controller for communications.

Figure 4.1
Simplified robot component skeleton

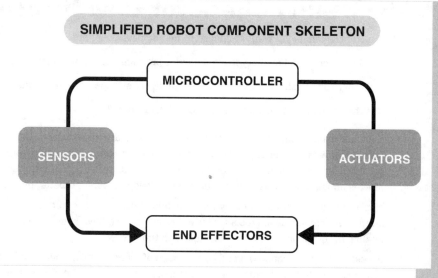

Figure 4.2
Basic layout of a microcontroller

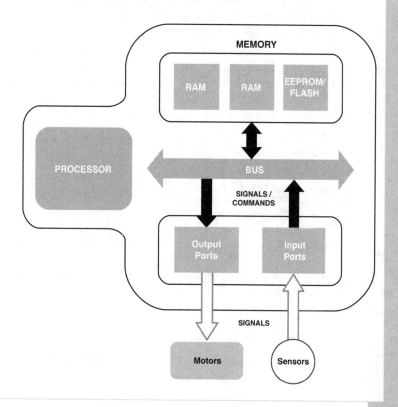

The Reality Check for the Microcontroller

Some of the things we want to check on the microcontroller are:

- How much actual memory is available for program use? (How much is actually free?)

- Is there any flash memory or access to auxiliary storage from the microcontroller?

- What is the data rate transfer on the microcontroller?

We don't want to have a situation where the sensors are generating more data or generating it faster than what the microprocessor and memory can handle. We don't want mismatches between sensor, motor, and microprocessor data transfer rates. Data acquisition rates versus data processing rates is an important area to consider when developing useful robot applications.

Microcontrollers typically come with some kind of utility that reports system information. The operating environment the microcontroller works in will also have some way of reporting system information. In this book, we get you started with the basics.

Accessing the information depends on the kind of robot the microcontroller is supporting. Some robots send feedback to the programmer via digital display monitor; others write the feedback to internal memory or some kind of external memory device accessible by the programmer. Let's take a look at two commonly found low-cost robot microcontrollers: the Arduino Uno and the Mindstorms EV3.

Figure 4.3 shows a photo of the Arduino Uno microcontroller we use to control one of the Arduino-based robot arms that we use for code examples in this book.

The Arduino ShowInfo is an example of the kind of utility that can be used with a microcontroller. The ShowInfo utility can be used to get the actual specifications of the microcontroller connected to an Arduino-based robot. Type the command, followed by the Enter key. Our version of ShowInfo presents the programmer with a menu like the one shown in Listing 4.1.

Listing 4.1 Arduino ShowInfo Menu

```
i    =    Show information
h    =    Show menu
r    =    Test serial communication
s    =    Speed tests
t    =    Timer Register Dump
?    =    Show menu
```

Examples of the output for the Speed tests option are shown in Listing 4.2 and Listing 4.3.

Listing 4.2 Example Output for the Speed Tests Option of the Arduino ShowInfo Utility

```
F_CPU = 16000000 Hz
1/F_CPU = 0.0625 us
```

Listing 4.3 is an example of the output for the next tests of runtime compensated for overhead. The interrupts are still enabled because millis() is used for timing.

Figure 4.3
Photo of our Arduino Uno microcontroller

ARDUINO UNO MICROCONTROLLER

Listing 4.3   Example Output for the Speed Tests of Runtime Compensated for Overhead

```
nop                  : 0.063 us
avr gcc I/O          : 0.125 us
Arduino digitalRead  : 3.585 us
Arduino digitalWrite : 5.092 us
pinMode              : 4.217 us
multiply byte        : 0.632 us
divide byte          : 5.412 us
add byte             : 0.569 us
multiply integer     : 1.387 us
divide integer       : 14.277 us
add integer          : 0.883 us
multiply long        : 6.100 us
divide long          : 38.687 us
add long             : 1.763 us
multiply float       : 7.110 us
divide float         : 79.962 us
add float            : 9.227 us
itoa()               : 13.397 us
```

```
ltoa()                     : 126.487 us
dtostrf()                  : 78.962 us
random()                   : 51.512 us
y |= (1<<x)                : 0.569 us
bitSet()                   : 0.569 us
analogRead()               : 111.987 us
analogWrite() PWM          : 11.732 us
delay(1)                   : 1006.987 us
delay(100)                 : 99999.984 us
delayMicroseconds(2)       : 0.506 us
delayMicroseconds(5)       : 3.587 us
delayMicroseconds(100)     : 99.087 us
```

The ShowInfo utility works for Arduino-based microcontrollers. Most microcontrollers come with their own version of a "show info" utility. For microcontrollers that have embedded Linux, there are additional options to find out information. For example, Mindstorms EV3 and WowWee's RS Media microcontrollers have embedded Linux and information on the processor, and opening proc/info can reveal some information about the processor. For example, the command:

more /proc/cpuinfo

on our EV3 microcontroller produces the output in Listing 4.4.

Listing 4.4 Example Output of more /proc/cpuinfo Command on the EV3 Microcontroller

```
Processor       : ARM926EJ-S rev 5 (v5l)
BogoMIPS        : 1495.04
Features        : swp half thumb fastmult edsp java
CPU implementer : 0x41
CPU architecture: 5TEJ
CPU variant     : 0x0
CPU part        : 0x926
CPU revision    : 5
Hardware        : MindStorms EV3
Revision        : 0000
Serial          : 0000000000000000
```

Executing the command uname on our EV3:

uname -a
LINUX EV3 2.6.33-rc4 #5 PREEMPT CET 2014 armv5tejl unknown

gives us the version of LINUX and processor for our EV3.

Running the free command on the EV3 controller gives us the output in Listing 4.5.

Listing 4.5 Example Output for Running `free` Command on EV3 Controller

```
root@EV3:~# free
              total        used        free      shared     buffers
    Mem:      60860       58520        2340           0        1468
   Swap:          0           0           0
  Total:      60860       58520        2340
```

These values give you some identifying information for components of the microcontroller, such as:

- Processor type
- Processor speed
- Memory capacity
- Memory free
- Memory in use

It's a good idea to compare information found running the microcontroller utilities or looking at `/proc/cpuinfo` and `/proc/meminfo` with the specification or data sheet that comes with the microcontroller and note any major discrepancies. Free memory limits the size of a program that can be loaded into the robot or the amount of sensor or communication data that can be handled in real-time by the robot.

Whereas the processor speed will limit how fast instructions can be executed by the robot and the data transfer rate between the processor and the ports, the available free memory has a major impact on how fast and how much sensor data can be processed. Keep in mind that the processor controls the flow of data and signals for the robot. So questions like:

- How much data can the robot handle?
- What's the largest chunk of data the robot can handle at one time?
- How many signals can the robot send or receive simultaneously?
- How fast can the robot process signals or data?

are all answered by the microcontroller's actual capability. The ATmega, ARM7, and ARM9 family of microcontrollers are the most popular and perhaps the most commonly found low-cost microcontrollers used for mobile and autonomous robots. All robots used for examples in this book use one or more of these microcontrollers.

Sensor Reality Check

Infrared sensors, ultrasonic sensors, light sensors, color sensors, heat sensors, and chemical sensors all sound like really cool sensors to have for a robot—and they are. But the devil is in the details. All sensors have some baseline capability, limit to their precision, and some amount of error relative to that baseline capability. All sensors have constraints on how much data they can gather, how fast they can gather it, and how precisely they can measure the data.

Ultrasonic sensors work by bouncing sound off objects and measuring the time it takes to receive the signal back, thus relating distance to time. But ultrasonic sensors differ in terms of how far they can actually measure distance, how accurately they can measure time, or how consistently they can measure time or distance.

Color sensors are a nice thing to have for a robot. Color can be used as a means of programming a robot when to take various actions and what actions to take. For example, if an object is blue, instruct the robot to do action A; if the object is red, do action B, wait until the blue object turns purple, and so on.

Let's say our robot has a color sensor that can distinguish between red, green, and blue. The issue is how accurate is the sensor. Which red does the sensor recognize? All shades of red? Can the sensor report the difference between various shades of blue—navy blue, sky blue, teal? Exactly how precise is the color sensor in distinguishing hues? If our robot is a bomb disposal robot and we instruct it to cut the orange wire and not the red wire, will there be a problem? Sensors measure things. Robots use sensors to measure things. The accuracy and limitations of the robot's sensors have to be considered when selecting a task for the robot to perform. Many sensors have to be calibrated on the first use. Some sensors have to be calibrated after each use. The calibration process is also an important part of considering what capability or limits a robot's sensors actually have.

BRON'S
Believe It
Or Not!

BRON'S BELIEVE OR NOT

The DARPA Robotic Challenge Finals were held June 5–6, 2015, at the Fairplex Fairgrounds in Pomona, California. Twenty-three robots competed in a series of challenges derived from search and rescue situations. The tasks to be performed were

- Drive and get out of a car
- Open a door and walk through the doorway
- Cut through a wall
- Turn a valve
- Perform a surprise task
- Walk through a debris field
- Walk on uneven terrain
- Climb stairs

There were 3 wheeled robot designs, one that was quadruped and wheeled design, and 18 bipedal robots. The robots ranged in cost between $500,000 and $4,000,000. Out of the 23 robots, only 3 successfully completed all the tasks. The winner was DRC-HUBO, followed by RUNNING MAN in second, and CHIMP in third place as shown in Figure 4.4.

Figure 4.4
The three 2015 Darpa Robotics Challenge Finalists with some brief information about each robot

2015 DARPA ROBOTICS CHALLENGE FINALISTS

TEAM KAIST

Robot Name: DRC-HUBO
Country: South Korea
Laboratories: HuboLab & RCV Lab
DOB: 2014

Height: 180 cm
Weight: 80 kg
Tasks Completed: 8
Time: 44:28

TEAM IHMC ROBOTICS

Robot Name: RUNNING MAN
Country : International
Institute: Institute of Human &
Machine Cognition
DOB: 2015

Height: 190 cm
Weight: 175 kg
Tasks Completed: 8
Time: 50:26

**TEAM TARTAN
RESCUE**

Robot Name: CHIMP
Country : USA
School: Carnegie Mellon
DOB: 2012

Height: 150 cm
Weight: 201 kg
Tasks Completed: 8
Time: 55:15

Although some of the robot designs allowed more autonomy than others, none of the robot designs were completely autonomous and all the robot designs required a significant amount of teleoperation and remote control. Even with significant amounts of teleoperation, many of the robots fell over at some point in the challenge.

Determine Your Robot's Sensor Limitations

Light visible to the human eye has a wavelength between 400 nm and 700 nm (see Figure 4.5). Wavelength is measured in nanometer, a unit of length in the metric system, equal to one billionth of a meter (0.000000001 m). Light that has a wavelength > 700 nm we call infrared. Light that has a wavelength < 400 nm we call ultraviolet. When a robot has an RGB (Red Green Blue) color sensor, that sensor measures lightwaves between 400 nm and 700 nm in length. For the human eye, every wavelength in that range is associated with a color.

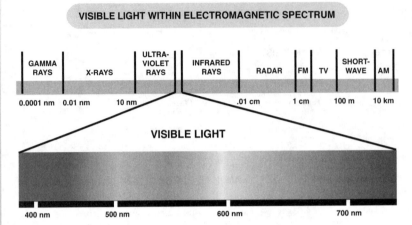

Figure 4.5
The wavelengths of light

The question is how precise and consistent is the resolution of a robot's color sensor? Does the sensor report that every wavelength light between 400 nm and 510 nm is blue without being able to distinguish between the different blues? Does it report that every wavelength between 570 nm and 700 nm is red and make no distinction between red and light red? How can we have 4-color RGB sensors and 16-color RGB sensors? What does that mean?

In Chapter 5, "A Close Look at Sensors," we look at how light and color sensors work and how robots use them to measure things, make decisions, and take action. Light sensors and color sensors aren't the only sensors that have these kinds of resolution questions.

Figure 4.6 shows how loudness, softness, low tones, and high tones of sound are measured. Some robots respond to loud sounds or soft sounds or to certain tones or kinds of sound.

Sound can be used as input to the robot's program. For example, if a sound is loud enough it causes the robot to take one action, and if the sound produces a certain tone it causes the robot to take some other action. The amplitude of the sound wavelength determines how loud or soft a sound is.

The frequency of the sound wavelength determines what tone the sound makes. So like the light sensor, a sound sensor measures wavelengths. Instead of light wavelengths, sound sensors measure sound wavelengths. The question is how loud is loud? Will the robot's sound sensor detect every sound above 50 dB (decibels) as loud and every sound below 50 dB as soft? Can the robot's sound sensor distinguish between 100 dB and 120 dB? It's not only sensor measurements of light and sound wavelengths that have these resolution questions.

Because most sensors measure analog values there will be questions of resolution. For instance, the acidity of a liquid is determined by a chemical sensor that measures the amount of hydronium ion present. Table 4.1 shows the pH scale.

Figure 4.6
Loudness, softness, low tones, and high tones of sound

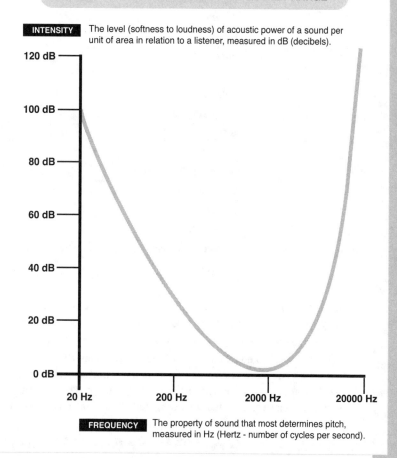

HUMAN HEARING INTENSITY AND FREQUENCY RANGE

INTENSITY The level (softness to loudness) of acoustic power of a sound per unit of area in relation to a listener, measured in dB (decibels).

FREQUENCY The property of sound that most determines pitch, measured in Hz (Hertz - number of cycles per second).

Table 4.1 The pH Measurement Scale

pH Value	Example of Solutions
14	Liquid drain cleaner
	Caustic soda
13	Bleaches
	Oven cleaner
12	Soapy water
11	Household ammonia (11.9)
10	Milk of Magnesia (10.5)
9	Toothpaste (9.9)

pH Value	Example of Solutions
8	Baking soda (8.4)
	Seawater
	Eggs
7	Pure water
6	Urine (6)
	Milk (6.8)
5	Acid rain (5.6)
	Black coffee (5)
4	Tomato juice (4.1)
3	Grapefruit & orange juice
	Soft drink
2	Lemon juice (2.3)
	Vinegar (2.9)
1	Hydrochloric acid secreted from stomach lining (1)
0	Battery acid

If our robot has a sensor that measures pH, would it be capable of distinguishing the difference between battery acid and lemon juice? Or would it report them both as an acid?

In our Robot Boot Camp Scenario #1, Midamba could use a robot to measure the difference between acid and bases to help him solve his battery problem. How much resolution does the robot's pH sensor need? The limits and resolution of all the robot's sensors should be identified and noted prior to picking situations, scenarios, and roles for the robot. The sensors roughly impact 25% of the robot's success in executing its tasks for assigned situations and scenarios.

Actuators End-Effectors Reality Check

A robotic arm is useful in placing, lifting, and positioning objects. But all robot arms have limits to how much weight or mass they can lift or hold, or how much torque can be generated by the servos on the arm. Not all robot arms are created equal, and not all robot arms can manipulate their maximum weight objects in all positions (even positions claimed by the specifications). Figure 4.7 shows a robot's arms with 2 and 6 DOF (Degrees of Freedom).

 note

Robot arm capability is often described as DOF (Degrees of Freedom). Simply put: How many modes or axes of movement can the robot arm be maneuvered in 3D space?

Figure 4.7
Robot arms with 2 and 6 DOF

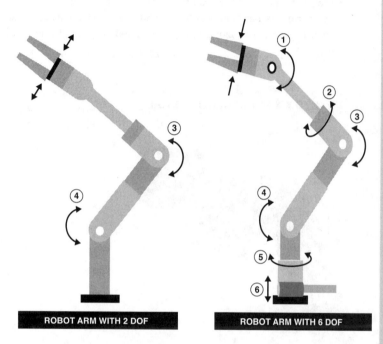

ROBOTS WITH 2 AND 6 DEGREES OF FREEDOM

ROBOT ARM WITH 2 DOF

ROBOT ARM WITH 6 DOF

1 Wrist Pitch 180°
2 Forearm Rotate Continuous 360°
3 Elbow Pivot 200°
4 Shoulder Pivot 200°
5 Shoulder Azimuth 300°
6 Shoulder Elevation

 note

Some code for the robot arm examples that we use in this book was executed on the Arduino-based PhantomX Pincher robot arm manufactured by Trossen Robotics. Table 4.2 lists the manufacturer's specifications for the arm.

Table 4.2 The Specifications for PhantomX Pincher Robot Arm

Specification	Limit
Vertical reach	51 cm
Horizontal reach	38 cm
Strength at reach	30 cm/200 g,20 cm/400 g,10 cm/600 g
Gripper strength	500 g holding
Wrist length strength	250 g, 150 g rotating

These are the kinds of specifications that need to be checked for robot arms. Sometimes the manufacturer's specifications are conservative and a component may have a little better performance than specified. For example, in this case our gripper strength was somewhat greater than 500 g, and rotating strength was a little more than 150 g. This is an important specification to note. The robot arm may be capable of holding 500 g but is not capable of rotating 500 g This fact will have an impact on how the robot arm is programmed and what it can and cannot do.

Figure 4.8 is a photo of the PhantomX Pincher robot arm used for the examples in this book.

PHANTOM X PINCHER ROBOT ARM

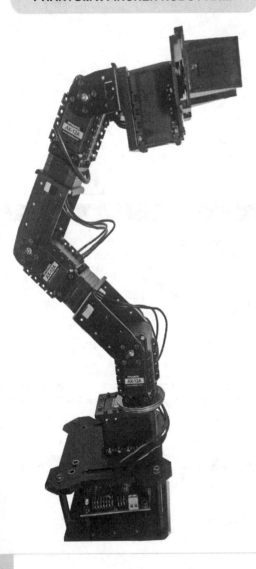

Figure 4.8
Photo of the PhantomX
Pincher robot arm

The robot capability matrix is a good place to record the actual capabilities. We use a spreadsheet to record our capability matrix. Table 4.3 is a sample of our actual robot capability matrix in a spreadsheet.

Table 4.3 A Sample of Our Robot Capability Matrix Spreadsheet

	A	B
1	**PhantomX Robot Arm**	**Limit**
2	DOF	5 d
3	Power requirements	12 v 5 amp
4	Vertical reach	51 cm
5	Horizontal reach	38 cm
6	Strength at reach	30 cm/200 g,20 cm/400 g,10 cm/600 g
7	Gripper strength	504 g
8	Wrist length strength	250 g, 150 g rotating
9		
10	**HCSR04 Ultrasonic Sensor**	
11	Effectual angle	< 15 degrees
12	Power requirements	5 v DC
13	Range	2 cm – 400 cm
14	Resolution	0.32 cm
15	Measuring angle	30 degree
16	Pulse width	10 uS

 note

Although we could put multiple robots on one spreadsheet, we keep it to one spreadsheet per robot, with all sensors, actuators, and end-effector capabilities for each robot recorded on separate spreadsheets. This keeps it easy when programming, and when looking up robot capabilities.

REQUIRE Robot Effectiveness

Recall from Chapter 1, "What Is a Robot Anyway?," we rated a robot's effectiveness in four areas:

- Sensor effectiveness

- Actuator effectiveness

- End-effector effectiveness

- Controller effectiveness

Each area is limited by the actual capabilities of the hardware specifications. According to our Robot Effectiveness Quotient, the

 note

REQUIRE stands for Robot Effectiveness Quotient Used in Real Environments. We use REQUIRE as an initial litmus test in determining what we can program a robot to do and what we won't be able to program that robot to do.

sensors make up 25% of the robot's effectiveness. The robot uses sensors to measure things and then take actions and make decisions based on those measurements. Table 4.4 shows some of the common kinds of quantities that robots use sensors to measure.

Table 4.4   Common Quantities Robot Sensors Measure

Quantity Type	Basic Unit/Symbol
Length	meter, m
Mass	kilogram, kg
Time	seconds, s
Temperature	Celsius, C
Electric Current	ampere, A
Area	m^2
Volume	m^3
Density	kg/m3
Speed	m/s
Acceleration	m/s^2
Force	newton, N
Pressure	pascal, Pa
Energy	joule, J

The quality, accuracy, precision, resolution, and limitations of the sensors should be determined prior to programming. For example, if a robot's sensor measures distance (length), then it has a maximum distance and a minimum distance that can be measured in terms of meters, centimeters, and so on. The ultrasonic sensors our robots use for the examples in this book have a maximum range of 100 cm and a minimum range of 10 cm. Objects outside that range cannot be detected by our robot's distance sensors. We include the notation for basic units and symbols in Table 4.4 because you need to be able to compare apples with apples when determining a sensor's limitation and precision.

Different sensor manufacturers specify sensor capability using different units of measure. It's a good idea to pick one standard of measurement and stick with it when programming robots. Numbers get thrown around a lot in a typical robot application. If the units of measure are mixed, then the program will be difficult to change, maintain, and reuse. For example, let's say our robot has both an infrared sensor and an ultrasonic sensor to measure distance simultaneously of two objects, and we want to know how far the objects are apart at any given time. We might have some robot code like the following:

```
Begin

    Object1Distance = Robot.ultrasonicSensorGetDistance();
    Object2Distance = Robot.infraRedSensorGetDistance();

    …
```

```
DistanceApart =   Object2Distance -  Object1Distance;
Robot.report(DistanceApart).
```

end

What does `DistanceApart` represent if the ultrasonic sensor uses inches and the infrared sensor uses centimeters? Sensors from different manufacturers may be on the same robot with Manufacturer #1 using yards for distance and Manufacturer #2 using meters to describe distance. Although it is not impossible to keep track of which is which, combining different units of measure complicates matters. Also, it is difficult to use the same program between different robots if the units of measure are not consistent between robots and libraries.

It is difficult to mix and match robot program libraries if each library uses a different standard of unit measurement. When instructing the robot to measure things, you should use the same standard of unit measurement for all instructions and objects that the robot interacts with. You should use the same unit of measurement standard for all robot instruction libraries. The robot capability matrix should include a sensor capability matrix with limits specified using the applicable standard unit of measurement. We clarify this idea as we explain how to develop ROLL models for your robots in Chapter 8, "Getting Started with Autonomy: Building Your Robot's Softbot Couterpart." In this book, we use SI (International System of Units) units when specifying the measurement of all quantities.

What's Ahead?

In Chapter 5, "A Close Look at Sensors," we discuss the different types of sensors and how they work for your robot. We discuss their capabilities and limitations.

5

A CLOSE LOOK AT SENSORS

Robot Sensitivity Training Lesson #5: *A robot can sense when you are mistaken.*

In Chapter 1, "What Is a Robot, Anyway?," we stated that the basic components of a robot are:

- One or more sensors

- One or more actuators

- One or more end-effectors/environmental effectors

- Controllers

Programming a robot's sensors, actuators, and end-effectors using a microcontroller is what programming the robot is all about.

The sensors, actuators, and end-effectors are what make a robot interesting, capable of performing tasks, and able to interact with its environment. Each sensor gives the robot some kind of feedback from the environment. But let's make sure we acknowledge the distinction between the "sense" and the "sensor."

A *sense* is a specialized function of a mechanism that receives and perceives the consequence of a stimulus on a sense component. A *sensor* is the sense component, or device that responds to the physical stimulus in some way. It is the device that detects or measures a physical property of the environment and then produces output as a reading, measurement, reaction, or signal that the stimulus was received.

Human sensors are our eyes, ears, skin, and so on. Robots have sensors such as cameras and ultrasonic and infrared sensors, and they have access to many different types of sensors, maybe hundreds. A single robot can be equipped with as many sensors as its real estate permits, its microcontrollers can connect, and its power supply can support. Table 5.1 shows a list of many different kinds of sensing, and the human and robot sensors.

Table 5.1 Human and Robot Sensors

Sense	Human	Robot
Sight	Eyes (color receptors, rods for brightness)	Camera, proximity (ultrasonic, etc.), color
Taste	Taste receptors (Tongue)	N/A
Smell	Smell receptors (Nose)	Gas sensor?
Tactile	Nerve endings	Touch sensor, artificial skin
Sound	Ear drums	Sound sensor and speakers
Nociception (pain)	Cutaneous (skin), somatic (bones and joints), and visceral (body organs)	N/A
Equilibrioception (balance)	Inner ears (vestibular labyrinthine system)	Gyroscope
Tension	Muscles	N/A
Thermoception (heat)	Hot/cold receptors	Barometer, temperature sensors, infrared thermometer
Magnetoreception	N/A	Magnetic sensor
Time	Cerebral cortex, cerebellum, and basal ganglia	Clock
Hunger	Ghrelin hormone	N/A
Thirst	Thirst receptors	N/A
Echolocation (navigation)	N/A	Ultrasonic, compass, GPS sensors
Electroreception (electric fields)	N/A	Electric field (EF) proximity sensors
Direction	Hippocampus and entorhinal cortex (EC)	Compass sensor
Proximity	N/A	Ultrasonic, EOPD, infrared sensors
Force, pressure	N/A	Force, pressure sensors

A robot equipped with many types of sensors and actuators can perform all kinds of physical behaviors. Robots can simulate all kinds of abilities by combining sensors with actuators and devices.

What Do Sensors Sense?

Some sensors are transducers, devices that convert one form of energy to another. Transducers are used to sense different forms of energy such as

- Magnetism

- Movement

- Force

- Electrical signals

- Radiant

- Thermal

Transducer types can be used for input or output conversions depending on the type of signal or process being sensed or manipulated. Sensors are *input devices* that change a physical quantity into its corresponding electrical signal that is then mapped to a measurement. The physical quantity is normally nonelectrical.

For example, a sound wave is a disturbance pattern caused by the movement of energy traveling through the air as it spreads out away from the source of the sound. A sound sensor is essentially a microphone that detects this form of energy.

 note

Many sound sensors are nothing more than a dynamic microphone that detects acoustic waves and then changes them to an electrical analog signal.

The electrical signal is generated by the diaphragm, a thin sheet of metal that collects sound waves causing the magnets that surround the diaphragm to vibrate. The vibration of the magnets causes a metal wire coil that surrounds the magnets to vibrate too, inducing a current in the coil that is converted to the electrical signal. The sensor then produces a signal that quantifies the loudness or softness of that original sound wave. Figure 5.1 shows the sound sensor's transformation of sound waves into an electrical signal measured in decibels.

Figure 5.1
The sounds sensor's transformation of sound waves

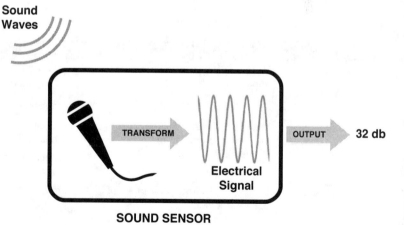

SOUND SENSOR WAVE TRANSFORMATION

Sound Waves

TRANSFORM

Electrical Signal

OUTPUT   32 db

SOUND SENSOR

Sensors measure the different forms of energy that exist in the robot's environment. Most of the time we are referring to the environment "outside" the robot, but there are sensors that measure the internal robot environment—the robot's *internal state*—as well. These sensors are called *proprioceptive*. Gyroscopes, accelerometers, and compass sensors are all examples of *proprioceptive sensors*.

 note

A gyroscope can calculate the changes in orientation and the rotational motion or angular velocity measured in RPM (rotations per minute or degrees per second). Accelerometers measure the robot's acceleration in units of meters per second squared or G-force. The compass sensor measures the earth's magnetic field and calculates a magnetic heading that reflects the direction the robot is facing.

Sensors that measure the external environment as it interacts, intersects, or affects the robot are called *exteroceptive* sensors. The robot is the point of reference for these types of contact, proximity, or ranging sensors.

Sensors that measure physical quantities in the environment such as surface temperature, pH levels in liquids, turbidity, air pressure, and magnetic fields are called *environmental* sensors. These quantities do not require the perspective of the robot. Figure 5.2 shows the different perspectives of the robot as it relates to proprioceptive, exteroceptive, proximity, and environmental sensors. Table 5.2 lists the types of sensors, a brief description, and examples.

note

Contact sensors are used to measure contact between the robot and some other object in the environment. Proximity sensors measure the distance to objects near (the robot) but not touching the robot sensor. Ultrasonic and optical sensors are examples of proximity sensors that use ultrasonic waves and light to measure the distance to an object.

Table 5.2 Types of Robot Sensors

Type	Description	Examples
Proprioceptive	Measures the robot's *internal state*	Gyroscope Accelerometer Compass
Exteroceptive	Measures the external environment as it interacts, intersects, or affects the robot	Proximity—Measures distance to object without touch (ultrasonic, optical) Contact—Measures contact between robot and an object (touch, pressure)
Environmental	Measures physical quantities in the environment	Temperature pH Turbidity Magnetic fields

Figure 5.2
Robot sensor perspectives

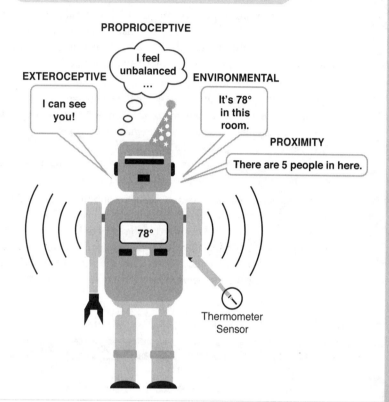

Analog and Digital Sensors

The type and the classification of a sensor gives you some insight into how the sensor works, what it measures, how it measures, and how it can be used. As transducers, sensors can be classified by the input signal that measures whether it is input from a proprioceptive, exteroceptive, or environmental sensor and by the output signals it produces.

 note

Analog and *digital* are the most basic sensor classifications that describe how a sensor measures. A digital sensor produces a noncontinuous discrete output signal, and an analog sensor creates a signal that has a continuous value. Analog sensors are meant to help convert real-world information that isn't electrical. Both types of sensors' output will be represented in a digital format recognizable by a processor. Some sensors do not *directly* turn their signals into digital signals but instead create an analog signal that is converted later.

Digital sensors produce values outputted as a single "bit" like a simple switch. When the switch is on, the circuit is closed and electricity flows through the circuit; when the switch is off, the circuit is open and electricity does not flow.

A switch is an example of a serial transmission that outputs a binary signal that has only two discrete states: 1 or 0 (ON or OFF). Bits can also be combined to produce a single byte output of n bits as a parallel transmission. An example of a digital sensor is the built-in optical incremental encoder. It outputs a relative position of the motor to the last position.

The encoder, coupled with a microcontroller and tachometer, measures the velocity and direction of the motor. The optical encoder detects the movement of the servo motor. We discuss the optical incremental encoders and servo motor in Chapter 6, "Programming the Robot's Sensors." A digital sensor also has embedded electronics, which directly process the signal inside the sensor. The data transmission is also digital, which means it is not sensitive to cable length, resistance, or impedance and is not influenced by electromagnetic noise. The digital sensor can also return multiple values that provide additional information about the reading.

Analog sensors simply produce the signal output, the voltage reading. That's all. True analog sensors do not have an embedded chip, so digital conversion is performed outside the sensor. Analog sensors are more accurate because the original signal is represented with a higher resolution. But these signals can be easily affected by noise or degrade in transmission. Analog signals are also hard to use in calculations and comparisons. But there is data loss when converting analog to discrete values. Table 5.3 compares some of the attributes of digital and analog sensors.

Table 5.3 Digital and Analog Sensor Attributes

Attribute	Analog Sensor	Digital Sensor
Type of signal	Continuous	Discrete
Accuracy of signal	High accuracy, close to the original signal	Some data loss
Signal conversion to digital	Loses some accuracy when converted to digital	No conversions
Usage of signal by microcontroller	Must be converted Difficult to be used in calculations	Ready to be used
Signal processing	Processing of signal outside sensor Requires amplification	Inboard electronics for processing No amplification required
Signal transmission	Sensitive to degrading and noise	No degradation during transmission
Signal output	Only voltage reading	May contain additional information

Reading Analog and Digital Signals

Both digital and analog sensors produce a signal, and that signal is mapped to or interpreted as a reading or measurement. But what is a signal? How can a signal be mapped to a measurement?

First, a signal is a time-varying quantity that conveys some kind of information. This means a signal is a value that changes over time; it is not constant. So the time-varying quantity is a voltage changing over time or a current. These signals can be passed through wires or through the air using radio waves as with WiFi or Bluetooth.

An analog sensor creates a signal that has a continuous value, and digital sensors produce a discrete output. With an analog sensor, the value it produces is proportionate to the quantity being measured. Examples of analog sensors are

- Temperature

- Pressure/force

- Speed/acceleration

- Sound

- Light

All these sensors measure analogous quantities that are naturally continuous. An analog signal can have a range of 0 V to 5 V (volts). No sound detected by a sound sensor would return an analog signal of 0 V, and the maximum sound detected would have a maximum of 5 V. Sound sensors detect signals of any value within that range. Figure 5.3 (a) shows an analog signal as a continuous signal between 0 V and 5V, continuously detecting changes in the signal over time. This signal is smooth and continuous. For the microprocessor in our robot controller to do anything with these readings such as compute values and perform comparisons between signals, the signal has to be converted to a digital signal by an *analog-to-digital* (A/D) converter. The A/D converter may reside on the microcontroller or be a part of the sensor. Converting the analog signal to discrete values is called *quantization* as shown in Figure 5.3 (b).

The A/D converter divides that range into discrete values where the maximum number of volts is divided by the n bits of the converter. For example, the Arduino A/D converter is a 10-bit converter. This means 10 bits represent each signal in the sample. There is a $2^{10} = 1023$ discrete analog level. The voltage resolution divides the overall voltage measurement by the number of discrete values:

$$5V/ 1023 = .00488 \text{ V} \approx 4.8 \text{ mV}$$

The voltage resolution is 4.8 mV (millivolts). Some A/D converters are 8 bits ($2^8 = 256$ discrete levels with a 19.4 mV), and others are 16-bit ADCs ($2^{16} = 65,535$ discrete levels with 0.076295109 mV). The voltage resolution is the difference between the two readings. The lower the difference the more accurate the readings. The more bits the better the digitization of the original signal and a lower error.

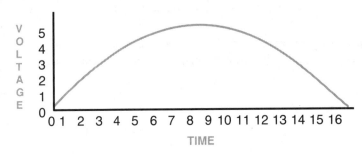

QUANTIZATION OF ANALOG SENSOR SIGNAL

(a) ANALOG SENSOR SIGNAL

Figure 5.3
(a) An analog signal as a continuous signal between 0V and 5V, and (b) the conversion of the analog signal into discrete values

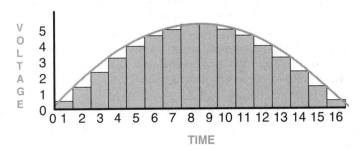

(b) QUANTIZATION OF SENSOR SIGNAL

Some digital sensors are really analog sensors with an A/D converter; true digital sensors generate a discrete signal. The sensor can output a range of values, but the value must increase in steps. Discrete signals typically have a stair-step appearance when they are graphed on chart as shown in Figure 5.4.

An example of a discrete sensor is a digital compass. It provides a current heading by sending a 9-bit value with a range from 0 to 359, that's 360 possibilities.

Figure 5.4
A graphed chart with discrete signals

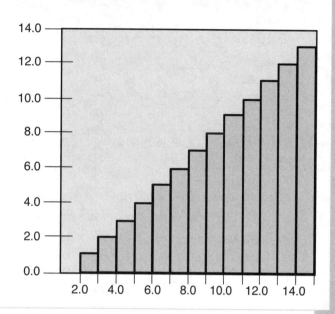

GRAPHED CHART WITH DIGITAL SIGNAL

The Output of a Sensor

An analog sensor produces an analog reading, which is the voltage of the signal converted to a digital value by the A/D converter. For example, what if a voltage reading was 4.38 V? What would the A/D converter return? To keep this simple and manageable, let's say we are using a 3-bit converter. For a 3-bit converter, there would only be 8 discrete levels or 8 possible digital values and the voltage resolution would be

5V / 8 = .625 V

So after the sensor converts the analog reading, it returns a digital output of 111. We can make a table of the analog readings; it's the binary representation in Table 5.4. As you can see, the ranges of analog signal readings are defined by the voltage resolution .625. The voltage reading of 4.38 V is in the range of 4.371 V to 5.00 V.

Table 5.4 Analog Readings and Their Binary Representation

Voltage Levels (V) (Voltage Resolution .625)	Binary Representation (3 bit)
0–0.62	000
0.621–1.25	001
1.251–1.87	010
1.871–2.5	011
2.51–3.12	100
3.121–3.75	101
3.751–4.37	110
4.371–5.00	111

Decimal values can be calculated:

8 / 5 V = ADC Reading / Analog measured value

In this case:

8 / 5 V = ADC Reading / 4.38 V

(8 / 5 V) * 4.38 V = ADC Reading

7.008 = ADC Reading

This is equivalent to the binary to decimal conversion:

111 ≈ 7.008

Decimal values are mapped to values (such as colors) or value ranges interpreted (such as bright and dark light).

Where Readings Are Stored

The measurements of the sensor may be stored in a data structure depending on the API being used. A data structure may be necessary if multiple values are returned. As mentioned earlier, digital sensors may return multiple values. For example, a sensor detecting the color of an object may return RGB values in a single structure. A sensor may also provide additional information about the reading. A sensor that is to detect the location of the object may also capture its distance, color, and so on. These structures may also be stored in an array if the sensor can make multiple readings. In Chapter 6 we discuss that some sensors, such as an ultrasonic, can take a single reading or perform continuously. It has a mode that allows for continuous readings. An advanced ultrasonic sensor stores the distance of multiple objects within its range. Each reading is stored in an array. We discuss simple data types used to store values in Chapter 6.

Active and Passive Sensors

Active and passive sensors describe how the sensors make measurements and how they meet their power requirements. Passive sensors receive energy from their environment or from the object to be measured. Passive sensors have no effect on their surroundings or whatever they are measuring. This is especially useful when robots are to be inconspicuous in a given scenario. Passive sensors are considered nonintrusive and energy efficient.

Active sensors involve direct interaction with their environment. They make observations by emitting energy into the environment and therefore require a power source. They are less energy efficient but are more robust because they are less affected by the available energy sources.

Let's contrast a passive infrared sensor (PIR) and an active ultrasonic sensor used for a robot object detection system. A PIR measures the infrared light that radiates off objects. The sensor detects when there is a change in the normal radiation. In Figure 5.5 (a), a robot enters the field of view of the sensor. The sensor is triggered because the robot breaks the continuous field. What the sensor detects is the disturbance. The sensor is passive because it does not emit an infrared beam; it just accepts the incoming radiation without any type of intervening.

Figure 5.5
(a) A passive infra-
red sensor for object
detection; (b) an ultra-
sonic sensor for object
detection

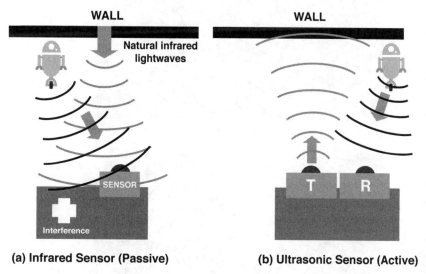

PASSIVE AND ACTIVE OBJECT DETECTION

WALL · Natural infrared lightwaves · SENSOR · Interference

WALL · T · R

(a) Infrared Sensor (Passive)   **(b) Ultrasonic Sensor (Active)**

An active sensor *generates* an electrical signal that changes the excitation signal by passing an electric current or pulse through the signal; the sensor measures the changes in the current that is reflected back. Using the active ultrasonic sensor for motion detection, sound waves are generated in the ultrasonic frequency range, typically 30 kilohertz to 50 kilohertz (kHz). The sensor emits cone-shaped sound waves in the 40 kHz frequency. This frequency is inaudible to people (which is 20 Hz to 20 kHz) and cannot pass through most objects.

The sensor listens to the sound that is reflected back off the objects in its field of view. The time it takes to transmit and then receive the reflected wave determines the distance to the object. So for a robot detection system, the ultrasonic sensor is continuously sending ultrasonic waves. If something is within the ultrasonic sensor's field of view, the sound waves are reflected off the new object, returning a different reading as shown in Figure 5.5 (b).

Some sensors have an active and passive mode or active and passive versions of the sensor. A digital camera is an example of a device that has both an active and a passive mode. Figure 5.6 shows two images of a group of tin toy robots: (a) is an image taken in passive mode; (b) is the same image taken in active mode. A digital camera has an image sensor that converts optical images to electronic signals, right? In passive mode, the sensor uses the existing ambient light. As you see, the robots' characteristics are not clear or illuminated in the low lighting. The image sensor records or captures the radiation provided in the environment.

PASSIVE AND ACTIVE DIGITAL CAMERA MODE LIGHTING ON ROBOTS

Figure 5.6
(a) Image taken in passive mode, (b) image taken in active mode

(a) A digital image of tin toy robots taken in **passive mode**.

(b) A digital image of tin toy robots taken in **active mode**.

If there is not enough light, the user can select a flash or the camera detects there is not enough light and switches to an active mode where a flash is used. The flash is its own energy source that illuminates the field and records the radiation that is reflected (Figure 5.6 (b)). Now the robots are better lit and the characteristics can be seen. This is also how active and passive modes work for a light sensor. The passive mode measures the ambient light rather than the reflected light from the LED in active mode. In active mode the sensor emits its own light source, the LED, and measures the light that is reflected off the objects.

Table 5.5 lists examples of active and passive sensors along with a description of each.

 note

There is an active version of the infrared sensor that works similarly to the ultrasonic sensor. It is the light version of the sound sensor. An active infrared sensor uses invisible light as opposed to high-frequency sound to scan an area. The light is reflected off the objects in the scan zone and detected by the receiver.

Table 5.5 Active and Passive Sensors

Sensor Type	Description	Examples
Active	Sensors that require an external power source; they generate an electrical signal that modifies the excitation signal and then measures the change of the current that is reflected back.	Ultrasonic GPS Color (active) Camera (with flash)
Passive	Sensors that do not require an external source of power; they convert the external stimulus's energy into the output signal.	Passive infrared Compass Electric field Temperature Chemical Touch Camera (no flash)

Sensor Interfacing with Microcontrollers

For these sensors to be of any use, they have to be connected to the microcontroller. There must be an interface for the sensors composed of an analog-to-digital convertor for analog sensors and a bus interface communication protocol that transfers data between components inside a computer (like SPI, UART, or I^2C). Sensors connect to the microcontroller via the *serial port,* the physical I/O connector that converts the data to a series of bits that flow over a single wire for transmission. A sensor plugs in to the serial port allowing for communication between the sensor and the microcontroller. Figure 5.7 shows three sensors plugged in to the EV3 microcontroller and sensors connected to the Arduino interface sensor shield used to connect magnetic and pH sensors.

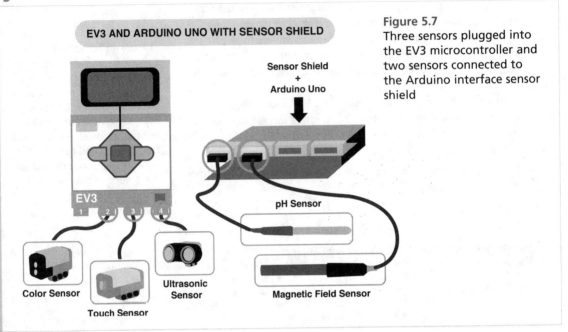

EV3 AND ARDUINO UNO WITH SENSOR SHIELD

Sensor Shield
+
Arduino Uno

EV3

pH Sensor

Color Sensor

Ultrasonic
Sensor

Magnetic Field Sensor

Touch Sensor

Figure 5.7
Three sensors plugged into the EV3 microcontroller and two sensors connected to the Arduino interface sensor shield

The microcontroller sends a signal to the sensor, and then the sensor sends a signal back to the microcontroller one bit at a time through the serial port. These signals are actually messages, and there are four types:

- System
- Command
- Info
- Data

The message is just part of the payload that may also include bits indicating the type of message and where the message starts or stops. The microcontroller can send a message composed of a command and info, and the sensor can send a message composed of data and info. Messages can be transferred through the serial port in two ways: *asynchronously* and *synchronously*. With each type of serial communication, there is a clock and a signal so communication can be coordinated. Each approach has a synchronization method that controls how and when the high or low, 1 or 0 bit is received.

With asynchronous data transfer, there is no common clock signal between the sender and receivers. The sender and the receiver agree on a data transfer speed (baud rate). The baud rate usually does not change until after the data transfer starts. Special bits are added to each word that are used to synchronize the sending and receiving units. There is a three-wire connection between the sensor and the microcontroller. The devices share a ground connection used as the reference point to measure voltage and two communication wires, one for transmitting data and the other for receiving data.

The *UART (Universal Asynchronous Receiver-Transmitter)* is an asynchronous serial communication. Your microcontroller and sensor may use this type of serial communication protocol (Arduino, Raspberry Pi, EV3, and a host of new EV3 sensors). Figure 5.8 shows a UART connection.

Figure 5.8
The three-wire UART connection

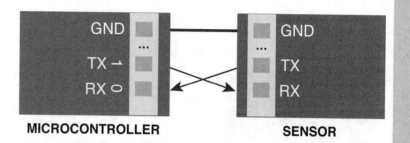

THREE-WIRE UART CONNECTION

UART has three wires: Tx is the transit wire, Rx is the receive wire, and GND is the ground wire. The transmit pin always transmits data, and the receive pin always receives data. Tx is connected to Rx, and Rx is connected to Tx.

I²C (Inter Integrated, I² part, Circuit) is a synchronous serial communication protocol to connect low-speed devices (like A/D converters, I/O interfaces, and sensors) to microcontrollers. The bus interface has an SDA wire used for a data signal, and the SCL wire is used for a clock signal.

The wires are bidirectional for sending and receiving data between sensors and devices. Devices on the I²C bus can be either masters or slaves. Masters initiate data transfers, and slaves react only to master requests. The current master dictates the speed of the clock that determines how fast the data will be transferred. But a slave device (sensor) can force the clock low at times to delay the master from sending data too fast or delay how fast it should prepare for transfer. There is no strict baud rate.

I²C is easy to use considering it can have more than one master or primary (usually the microcontroller) and an almost infinite number of slaves or secondaries (devices and/or sensors) utilizing the same bus by using an I/O/ port expander. Figure 5.9 shows the bus interface with one master and multiple slave devices.

> **note**
>
> With synchronous data transfer, both the sender and the receiver access the data according to the same clock with a special line for the clock signal. The microcontroller provides the clock signal to all the receivers in the synchronous data transfer. There is another line or wire for data transmission. When the clock pulses, a bit of data is transmitted. The sensors are made aware as to when to listen to the bits coming down the wire and when to ignore them.

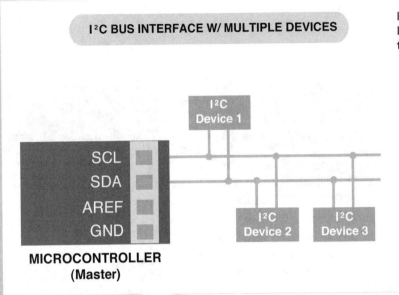

I²C BUS INTERFACE W/ MULTIPLE DEVICES

Figure 5.9
I²C bus interface with multiple slave devices

As mentioned earlier, a single robot can be equipped with as many sensors as its real estate permits, its microcontrollers can connect, and its power supply can support. This is one way it's done. Using the I²C bus interface, more than 1,000 devices can be connected to the microcontroller. Table 5.6 shows some of the attributes of UART and I²C serial bus interfaces.

Table 5.6 UART and I²C Serial Bus Interface Attributes

Attribute	UART	I²C
Communication type	Asynchronous	Synchronous
Typical use	Keyboard, character LCD/monitor	Multiple devices on a common bus
Typical speed	9 kHz to 56 kHz	Standard mode: 100 Kbps
	Varying	Full speed: 400 Kbps
		Fast mode: 1 Kbps
		High speed: 3,2 Mbps
Baud rate	Set between devices; cannot change until after data transfer	Can vary during data transfer
Number of devices on bus	One-to-one communication	Multiple devices (one master at a time)
Communication	One way	Bidirectional
Wires	GND - Ground	GND - Ground
	Tx - Transmit	SCL - Clock
	Rx - Receiver	VCC - Power
		SDA - Data

Attributes of Sensors

The attributes or characteristics of sensors describe things like the reading ranges, how long it takes the sensor to respond to a stimulus, its overall accuracy, and so on. What is its resolution and repeatability? With several sensors connected to the microcontroller, the voltage level and the power consumption have to be considered. That may determine how long an individual sensor is used or how long other sensors with a high power consumption are used.

Table 5.7, for example, compares compass sensors. The resolution could depend on the version of firmware in use. Comparing these characteristics helps in deciding which sensor will work the best for a robot scenario.

 tip

Being aware of these attributes helps determine the sensor limitations and which sensors can be used to compensate for those limitations. These attributes can also assist in comparing the quality of sensors by different manufacturers.

Table 5.7 Compass Sensor Comparison

Manufacturer	Resolution	Refresh Rate	Range	Multiple Readings
HiTechnic	1°	100 x per sec.	0-359	N/A
Mindsensors	.01° NBC, RobotC 1.44° NXT-G	N/A	0-359	Byte, Int, Float

Sensors have a number of attributes listed in Table 5.8. The values for these attributes should be supplied by the manufacturer of the sensor as part of the documentation or datasheet, but not all manufacturers supply this level of technical information. There may be other information manufacturers feel is important to their customers. Most manufacturers supply accuracy, response time, range, and resolution, and they believe linearity and repeatability are not so important. But importance is relative and depends solely on the system and the scenario in which the sensors are used.

Table 5.8 Sensor Attributes

Characteristic	Description
Resolution	The smallest change of input that can be detected in the output; it can be expressed as a proportion of the reading or in absolute terms.
Range	The maximum and minimum values that can be measured.
Linearity	The extent to which the actual measurement of a sensor departs from the ideal measurement.
Accuracy	The maximum difference that exists between the actual value and a new reading of the sensor; it can be expressed either as a percentage or in absolute terms. To calculate: 1 - ((Actual Value - Expected Value) / Expected Value)
Response time	Time required for change from its previous state to a final settled value.
Refresh rate	How often the sensor takes a reading.

Characteristic	Description
Sensitivity	Change required to produce a standardized output change.
Precision	The capacity of a sensor to give the same reading when repetitively measuring the same quantity under the same prescribed conditions; NOT closeness to the true value; related to the variance of a set of measurements; precision is a necessary but not sufficient condition for accuracy.
Reliability	Repeatability and consistency of the sensor.
Repeatability	The capability of a sensor to repeat a measurement when put back in the same environment.
Dimension/weight	The size and weight of the sensor.

Range and Resolution

Range and resolution are two of the most common attributes that people are interested in.

Range is the difference between the smallest and the largest outputs that a sensor can produce or the inputs in which the sensor can properly operate. These values can be absolute or a percent of the appropriate measurement.

For example, let's compare the ranges for a few sensors. For the ultrasonic sensor, the output range is 0 cm to 255 cm because distance is being measured, and the output range for a compass is 0° to 360°. But for a light sensor, 0 to 1023 is returned to the microcontroller.

In a well-lit room, the darkest reading may be 478 and the brightest reading may be 891. In a dimly lit room, the darkest reading may be 194 and the brightest reading might be 445. So a scale is used where 0 is dark and 100 is brightest represented as a percentage. The sensor is calibrated to the light of that particular environment. So from the example, in the well-lit room, once calibrated, 478 will have a reading of 0, and 891 will have a reading of 100.

Resolution is the minimum step size within the measurement range of the sensor. For example, the resolution of the HiTechnic compass was 1° where the resolution of an ultrasonic sensor is 3 cm. The resolution affects the accuracy of the sensor.

Remember the voltage resolution? With a voltage resolution of .625, an accurate reading of .630, the decimal representation is closer to 000 than it is to 001. For an ultrasonic sensor, an accurate reading of 7.5 cm could not be represented.

Precision and Accuracy

Precision is the capacity of a sensor to give the same reading when repetitively measuring the same quantity under the same conditions. Precision implies agreement between successive readings and *not* the closeness to the true value, which is accuracy. *Accuracy* is the maximum difference that exists between the actual value and a new reading of the sensor. The difference between the true value and the new reading is the absolute or relative error:

Absolute Error = New Reading − True Value

Relative Error = Absolute Error / True Value

note
The smaller the resolution the more accurate the sensor.

Precision and accuracy are actually unrelated to each other, meaning a sensor can be *precise* but not *accurate*. Precision is also as a synonym for the resolution of the measurement; for example, a measurement that can distinguish the difference between 0.01 and 0.02 is more precise (has a greater resolution) than one that can only tell the difference between 0.1 and 0.2 even though they may be equally accurate. So precision and resolution are also frequently abused. Figure 5.10 shows the various relationships between precision and accuracy.

Figure 5.10
Relationship between precision and accuracy

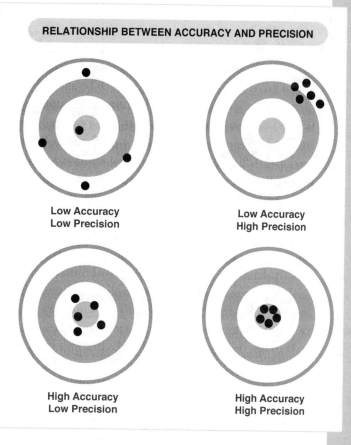

RELATIONSHIP BETWEEN ACCURACY AND PRECISION

Low Accuracy
Low Precision

Low Accuracy
High Precision

High Accuracy
Low Precision

High Accuracy
High Precision

Linearity

Linearity is the relationship between the input and output variations in analog sensor readings. Linearity can be used to predict the readings of the sensor (based on input), used to determine accuracy and measure error. If a sensor's output is linear, any change in the input at any point within its range produces the same change in the output. The output is proportional to input over its entire range. The graph of the slope of output versus input would be a straight line.

For example, if the ratio of input to output is 1 to 1, if there is an increase in the input (the stimulus) of the sensor by two, that would be reflected in the output. Ideally, sensors are designed to be linear, but not all sensors are linear when it comes to actual values. Figure 5.11 (a) shows the ideal linear relationship between input and output and the measured curve of a fictitious sensor and where the maximum error occurs. Figure 5.11 (b) shows the linearity of an ultrasonic sensor.

MAXIMUM ERROR AND LINEARITY OF US SENSOR

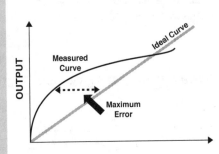

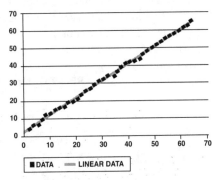

(a) Ideal linear relationship between input and output, and where the max. error occurs for a fictitious sensor.

(b) The linearity of an ultrasonic sensor.

Figure 5.11
(a) Ideal linear relationship between input and output, and where the maximum error occurs for a fictitious sensor; (b) the linearity of an ultrasonic sensor

When measuring the distance to an object 2 cm to 64 cm from the sensor, using 63 samples, the average error was 1.079 cm. Often nonlinearity is specified by a percentage:

Nonlinearity (%) = ($D_{in(max)}$ / $In_{f.s.}$) * 100

where $D_{in(max)}$ is the maximum input deviation, and $IN_{f.s.}$ is the maximum full-scale (f.s.) input. But the linearity of a sensor, like the ultrasonic sensor, depends on the conditions in which the readings were taken. A sensor should have linear readings in the best conditions. But if an environment is hostile to the performance of the sensor (for example, if there were multiple ultrasonic sensors transmitting sound waves at the same time) or the target object is not ideal (oddly shaped or odd position), the performance of the sensor will degrade and readings will not be linear.

Sensor Calibration

As mentioned earlier in this chapter, programming the robot's sensors, actuators, and end-effectors using a microcontroller is what programming the robot is all about. The sensors along with the actuators and end-effectors are what make a robot interesting, capable of performing tasks, and able to interact with its environment.

We have discussed many types of sensors, what they measure, how they measure, and how well they measure. There are a lot of good sensors available to be used by robot enthusiasts to program an exceptional robotics system. But sensors are not perfect. And as we discuss actuators in Chapter 7, "Programming Motors and Servos," we see that actuators are not perfect either. Many sensors are good enough right out of the box for many noncritical applications, but to achieve the best possible accuracy, precision, and so on, a sensor should be calibrated in the system where it will be used. There are several reasons why sensors may not perform as expected. Table 5.9 lists some of the issues.

Table 5.9 Factors That Warrant Sensor Calibration

Factor	Description
Errors in manufacturing	Inconsistent manufacturing
Exposure to varying temperatures	Sensitivity to heat, cold, shock, humidity during storage, shipping, or assembly
Worn over time	Parts become worn over time due to usage
Reliability on other components	Other components no longer reliable, which affects measurements

 note

What makes a sensor good is when a sensor has ideal values for all the attributes discussed earlier. When referring to accuracy, it is really a combination of precision, resolution, and calibration. *Calibration* is the method of improving the sensor's performance by removing structural errors in the sensor readings or measurements. *Structural errors* are differences between a sensor's expected output and its measured output.

These errors show up consistently every time a new measurement is taken. But any of these errors that are repeatable can be calculated during calibration, so that during actual use the measurements made by the sensor can be compensated in real-time to digitally remove any errors. If you have a sensor that gives you repeatable measurements with good resolution, it can be calibrated for accuracy.

Problems with Sensors

Sensors are manufactured devices, and two sensors from the same manufacturer may produce some slightly different readings due to errors in production. Sensors are also sensitive to environmental changes, such as heat, cold, shock, humidity, and so on, they may have been exposed to during storage, shipment, or assembly. This may ultimately show up as a change in a sensor's responses. Some sensors can actually become worn over time, and their responses naturally change over time requiring periodic recalibration.

Also keep in mind that the sensor may be only one component of the robot's measurement system. For example, with analog sensors, the ADC is also part of the measurement and is subject to variability. Light and color sensors can be affected by spectral distribution, ambient light, specular reflections, and the effectiveness of the LED. These factors are discussed in Chapter 6 in the section on programming color sensors.

End User Calibration Process

The manufacturer performs calibration on the sensor, of course. But as mentioned earlier, recalibration may be necessary. This is done by the end user to improve sensor measurement accuracy. A known value or way to accurately take a measurement is needed to perform the calibration. Table 5.10 lists the possible sources that can provide a standard reference to calibrate against.

Table 5.10 Standard References for Calibration

Standard Reference	Description
Calibrated sensor	A sensor or instrument that is known to be accurate can be used to make reference readings for comparison.
Standard physical reference	Reasonably accurate physical standards used as standard references for some types of sensors. For example: Rangefinders: Rulers and meter sticks. Temperature sensors: Boiling water - 100°C at sea level, Ice-water Bath - The "Triple Point" of water is 0.01°C at sea level. Accelerometers: Gravity is a constant 1G on the surface of the earth.

Each sensor has a "characteristic curve." This curve defines the sensor's response to an input. The calibration process maps the sensor's response to an ideal linear response. The best way to do this depends on the nature of the characteristic curve. If the characteristic curve is a simple *offset*, this means that the sensor output is higher or lower than the ideal output. Offsets are easy to correct with a single-point calibration. If the difference is a slope, this means that the sensor output changes at a different rate than the ideal. The two-point calibration process can correct differences in slope. Very few sensors have a completely linear characteristic curve. Some sensors are linear enough over the measurement range. But some sensors need complex calculations to linearize the output.

Calibration Methods

In this section we discuss two calibration methods:

- One-point calibration
- Two-point calibration

One-point calibration is the simplest kind of calibration and can be used as a "drift check" to detect changes in a response or when there is a deterioration of the performance of a sensor. One point can be used to correct sensor offset errors in these cases:

- **When only one measurement point is needed:** If the sensor usage only needs a single-level accurate measurement, and there is no need to worry about the rest of the measurement range. For example, a robot is using an ultrasonic sensor to position itself 3 cm from a container.

- **When the sensor is known to be linear and have the correct slope over the desired measurement range:** Calibrate one point in the measurement range and adjust the offset if it's necessary.

To perform a one-point calibration, follow these steps:

1. Take a measurement with your sensor.

2. Compare the measurement with your reference standard.

3. Subtract the sensor reading from the reference reading to get the offset.

   Offset = Reference_reading – Sensor_Reading

4. In the program, add the offset to every sensor reading to obtain the calibrated value.

For the example mentioned, to calibrate the ultrasonic sensor use a measuring tape for the reference standard. Position the robot exactly 3 cm from the container. Take a reading with the sensor. If the reading is not 3 cm but is 4 cm, then there is a –1 cm offset. Subtract 1 cm from every reading. This will work if the sensor has a linear characteristic curve and will be accurate over most of its range.

 note

We used sensors from Venier, HiTechnic, WowWee, and the LEGO Mindstorms Robotic kit for this book. Although these are not industrial-strength sensors, they are pretty good for what we used them for. They are low cost and easy to use. Table 5.11 lists the sensors we used and their sources.

Table 5.11 Sensors Used and Their Sources

Type	Description	Sensor	Manufacturer
Proximity and presence	Measures the distance the robot sensor is from an object using sound waves	Ultrasonic	EV3 Mindstorms
Heading	Measures the orientation of the robot in relation to a fixed point	Compass	HiTechnic
Imaging, color, light	Collects data from the surface of objects	Color	HiTechnic
		Camera	Charmed Labs
		RS Media Robosapien Camera	WowWee
Environmental	Measures physical quantities in the environment	pH	Vernier
		Magnetic field	Vernier

What's Ahead?

In this chapter, we talked about the different types of sensors, how they work, and what they measure. In Chapter 6, we will discuss how to use and program sensors, namely the color, Pixy Vision, ultrasonic, and compass sensors.

6

PROGRAMMING THE ROBOT'S SENSORS

Robot Sensitivity Training Lesson #5

A robot can sense when you are mistaken.

For a mobile robot, having some type of vision is critical. A mobile robot needs to avoid obstacles; determine its location and the distance to objects; and detect, recognize, and track objects using a multitude of sensors such as infrared, ultrasonic, cameras, image sensor, light, color, and so on.

Robot vision is a complicated topic and beyond the scope of this book. But a robot may not need a full vision system; it may require only a few of these capabilities based on the scenario and situation it is programmed for. In this chapter we focus on programming sensors that play some part in the robot's capability to "see"—that is, *to perceive or detect as if by sight.*

For a robot to see it uses

- Color/light sensors

- Ultrasonic/infrared sensors

- Camera, image sensors

> **note**
>
> We talked about senses and sensors at the beginning of Chapter 5, "A Close Look at Sensors." We made the distinction between the sense, sight, and the organ (eyes for humans and a bunch of sensors for robots).

Table 6.1 lists these sensors and how they assist in the robot's capability to see.

Table 6.1 Sensors That Assist the Robot to "See"

Sensor	Description
Color/light	Detects the color of an object; senses light, dark, and intensity of light in a room
Ultrasonic/infrared	Proximity sensors used to measure the distance from the sensor to an object in its field of view using sounds waves reflected off objects; infrared (IR) sensors measure light radiating from the object
Camera/image	Digital camera used to capture an image of the environment that can then be processed to identify objects

Using the Color Sensor

A color sensor detects the color of objects. It is made of two simple components: an LED and a photo-resistor that is sensitive to light. The LED projects a light beam onto a target object, and the photo-resistor measures the reflected light off the object. This is called *reflective color sensing*. When the light beams hit the target (incident light), three things can happen to incident light:

- **Reflection:** A portion is reflected off the surface of the target.

- **Absorption:** A portion is absorbed by the target surface.

- **Transmission:** A portion is transmitted.

The amount of reflection and absorption is based on properties of the object. The reflected light is what is sensed.

There are two types of reflection—diffuse and specular—and both contribute to what the detector senses. The diffuse reflection carries the most useful information about the surface color, but varying specular reflection can negatively affect the performance of the sensor.

Mostly specular reflection is almost constant for a given target and could be factored out from the readings. The material of the object determines the amount of diffuse and specular reflectance. A matte surface has more diffuse than specular reflectance, and a glossy finish has more specular reflectance. Figure 6.1 (a) shows incident light (from the LED), diffuse, and specular reflections, (Reflection, absorption, and transmission are all identified.) The intensity of the reflected wavelengths is analyzed to figure out the surface color of the object. For example, if a color sensor emits a red LED on a ball, how is color detected? The sensor actually records the intensity reflected off the object. So, if the ball is red, the ball reflects back high intensity of red indicating the surface is in fact red. If the ball is blue, the red reflected back would not be very intense because most of the light would be absorbed by the object. The object is not red. (see Figure 6.1 [b]). But what if the ball is not a primary color (red, blue, green)?

Figure 6.1
(a) The incident light (from the LED), diffuse and specular reflections, (b) shows how color of object is detected based on specular light

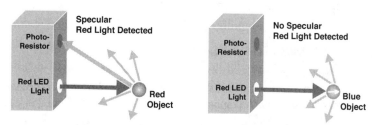

COLOR DETECTION

(a) Incident Light, Diffuse Light, and Specular Light

(b) With a **RED** LED light and a **RED** object, specular light is detected.

With a **RED** LED light and a **BLUE** object, light is absorbed and diffused. Little or no specular light is detected.

What if the color is orange or yellow; how is the color detected? An orange ball is closer to red in the color spectrum than the yellow ball, so some red light would be reflected back. So for all the colors the sensor is designed to detect, it would report the color closest to it theoretically. The sensor may not be able to detect these colors at all returning an incorrect value. On the less sensitive end, color sensors detect only four or six colors; others can detect 16 or 18 colors under good lighting conditions. Figure 6.2 shows what colors are detected based on a color spectrum.

A typical color sensor has a high-intensity white LED that projects modulated light on the target object. Other sensors have RGB LEDs. With the white LED, the reflection from the object is analyzed for the constituent red, green, and blue (RGB) components and intensities. In memory, the sensor has all the ranges of RGB for all the colors it can recognize. The level of reflected light is compared to the value stored in the sensor's memory. If the value is within a specified range, the sensor recognizes that color.

Color sensors combine the signal over the entire area of the light spot on the target object or *field of view* (FOV). So, if within the FOV there are two colors, the sensor sees the combination of the colors rather than each separate color. Color normalization removes all intensity values from the image while preserving color values. This has the effect of removing shadows or lighting changes on the same color pixels.

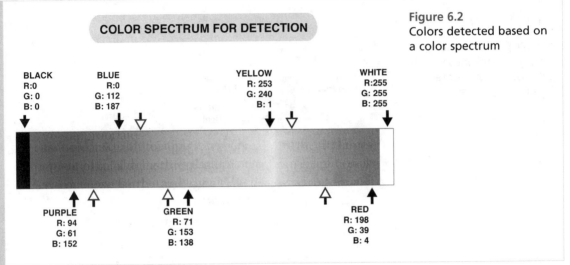

Figure 6.2
Colors detected based on a color spectrum

Color Sensor Modes

Color sensors can be set for different modes, and these modes report different aspects of the reading. For certain modes, the LED may or may not be used. Modes are made available via the API being used. Some examples of color sensor modes, described in Table 6.2, are

- Color ID
- RGB
- Ambient light level
- Light intensity level
- Reflected light intensity
- Calibration minimum and maximum

Table 6.2 Some Color Sensor Modes and Their Output

Color Sensor Mode	Description
Color ID	Color ID number.
Red	With the use of a red LED, intensity level of reflected red light.
Normalized RGB	Intensity levels of RGB as a normalized value.
Component RGB	Intensity levels of RGB as a separate value.
Ambient light level	Intensity levels of ambient light as a normalized value. (LED is off.)
Reflected intensity level	Intensity levels of reflected as a normalized value.
Calibration minimum and maximum	Specifies the maximum and minimum light intensities. After this calibration, maximum light is reported as 100 (or the highest value used) and minimum light is reported as 0.

Detection Range

The object should be as close as possible to the sensor to detect its color. Many recommend that the sensor be placed close at an angle over the target object so the reflected LED does not affect the photo-resistor. This is true for inexpensive sensors and may not be practical for a given scenario for the robot though.

The sensor takes an average of the entire FOV, so for the most precise results it is good to keep this field of view minimized.

A far-away reading may be desirable, for example, tracking the color of an object as the robot and color sensor move closer to the object. It may be necessary to track the color changes, for example, from a white color to an FOV that is an average of two or more colors and then back to a solid color as its target. The detection range should be specified in the manufacturer's datasheet of the sensor. Again, even with the color sensor, the range is affected by the use of the sensor in the application.

 note

Sensors have a minimum and a maximum range for accurate readings, but sometimes manufacturers do not list the maximum range.

 tip

Remember, as the object is farther away from the LED, the FOV increases in size.

Lighting in the Robot's Environment

What attributes of the environment can prevent the robot's sensors from being accurate? Ambient light is one of the biggest factors that can sabotage color readings. Ambient light includes any additional light in the environment that can affect the way a color appears. It's always present, and it can change the readings the robot receives at 8 a.m. compared to 8 p.m., or on one side of the room compared to the other side of the room.

The reflected light from the LED affecting the photo-resistor is considered ambient light as well. Shielding the sensor is a way to protect the photo-resistor from ambient light and is commonly used on surveillance cameras. Shielding refers to surrounding the sensors with something to prevent ambient light from interfering with the readings. This is especially valuable if the robot's environment has drastically different light settings (from well-lit to very dark).

⚠ caution

It's of the utmost importance that ambient light does not affect the quality of the sensors, or readings can end up being wildly divergent and inaccurate resulting in inconsistent and inferior results.

 tip

The effect of ambient light can be lessened by using calibrated values. These values can be considered a baseline for accurate readings. Errors or offsets can be calculated and used to increase the accuracy of new readings.

Calibrating the Color Sensor

The color sensor should be calibrated for a specific application to ensure that it provides the expected readings. Before using the sensor in the application, detect the color of a range of target object(s), record readings, and then make a chart of this data. So when sensing the same object (within the system), compare the new reading with the calibrated reading. Some APIs for sensors define methods to calibrate the color sensor. The color sensor is calibrated for the minimum and maximum, black and white, for a specific application.

Correcting errors in sensor readings may have to be calculated wherever a reading is made. Say that the birthday party is in a room with a lot of natural light on a partly cloudy day. That means clouds blowing over the sun can change the light in the environment while the party is going on. BR-1 must distinguish the cake plates on the table from the color of the tablecloth on the table. During the calibration phase, BR-1 measures an analog value of 98 for the gray tablecloth, 112 for the white plate, and then stores these values in memory. Now BR-1 is in use and has to detect the white plate, but the sensor readings are slightly different because of the lighting. A 108 not 112 reading for the white plate and 95 not 98 for the value of the tablecloth are measured. What does that mean? Is it detecting the plate?

Using the *thresholding method*, you add both calibrated numbers and then divide by 2 to find the average number. For example:

(98 + 112)/2 = 105 (threshold)

Anything above the threshold would be the white plate, and anything under would be the gray tablecloth.

If there three or four colors have to be distinguished, how is a threshold figured out? In that case, a *similarity matching* method can be used. The question becomes, how similar is each color to the calibrated value? Using similarity matching, calculate:

similarity = abs(new_reading − calibrating_reading)/calibrated_reading * 100

So:
gray tablecloth = (95 − 98)/98 * 100 = 3.06% difference
white plate = (108 − 112)/112 * 100= 3.5% difference

Since the 3.06% < 3.5%, making the gray tablecloth reading more similar to the calibrated value than the white plate, the BR-1 sees the tablecloth. This method can be used for any color and any number of colors, given calibration values are calculated beforehand. Look at color calibration as a way of showing the robot the difference between colors.

Programming the Color Sensor

In this section, we show BURT Translations for programming the HiTechnic color sensor using the EV3 microcontroller and the color sensor for the RS Media Robosapien. Listing 6.1 is the BURT Translation for programming the HiTechnic color sensor using the leJOS API `HiTechnicColorSensor` class. Burt's Translation Listing 6.1 shows the softbot frame for detecting a color and the Java code translation. This is the main line to test the color sensor and perform some basic operations.

BURT's Translation Listing 6.1   Unit2's Color Sensor Test Softbot Frame

BURT Translation INPUT

```
Softbot  Frame
Name:  Unit2
Parts:
Sensor  Section:
Color Sensor


Actions:
Step 1: Initialize and perform any necessary calibration to the color sensor
Step 2: Test the color sensor
Step 3: Report the detected color, modes, sample size and content


Tasks:
Test the color sensor and perform some basic operations.
End Frame
```

BURT Translations Output: Java Implementations

```
32    public static void main(String [] args)  throws Exception
33    {
34        softbot5 SensorRobot = new softbot5();
35        SensorRobot.testColor();
36        SensorRobot.closeLog();
37    }
38
```

In line 34, the softbot SensorRobot is declared. Three functions/methods are then called. We discuss the constructor and SensorRobot.testColor(). Listing 6.2 shows the code for the constructor.

Listing 6.2   The Constructor for the SensorRobot Object

```
1    public softbot5() throws Exception
2    {
3        Log = new PrintWriter("Softbot5.log");
4        ColorVision = new HiTechnicColorSensor(SensorPort.S2);
5    }
```

In line 4, a HiTechnicColorSensor object is declared. SensorPort.S2 is the serial port used for this sensor. The BURT Translation in Listing 6.3 shows the testColor()softbot frame and Java implementation.

BURT Translation Listing 6.3   The `testColor()softbot` **Frame and Java Implementation**

BURT Translation INPUT

```
Softbot Frame
Name: Unit2
Parts:
Sensor   Section:
Color Sensor

Actions:
Step 1: Test the color sensor by reporting mode name, color ID, mode,
        name, and RGB mode and name.
Step 2: Get and report sample size and content of the sample.

Tasks:
Test the color sensor and perform some basic operations.

End Frame
```

BURT Translations Output: Java Implementations

```
6
7     public void testColor() throws Exception
8     {
9         Log.println("Color Identified");
10        Log.println("ColorID Mode name =
                   " +        ColorVision.getColorIDMode().getName());
11        Log.println("color ID number = " + ColorVision.getColorID());
12        Log.println("Mode name = " + ColorVision.getName());
13        Log.println(" ");
14        Log.println("RGB Mode name =
                   " + ColorVision.getRGBMode().getName());
15        Log.println("RGB name = " + ColorVision.getName());
16
17        float X[] = new float[ColorVision.sampleSize()];
18        Log.println("sample size = " + ColorVision.sampleSize());
19        ColorVision.fetchSample(X,0);
20        for(int N = 0; N < ColorVision.sampleSize();N++)
21        {
22            Float Temp = new Float(X[N]);
23            Log.println("color sample value = " + Temp);
24        }
25    }
```

Here is the output for Listing 6.3:

```
Color Identified
ColorID Mode name = ColorID
color ID number = 2
Mode name = ColorID
RGB Mode name = RGB
color ID number = 2
RGB name = ColorID
sample size = 1
color sample value = 2.0
```

The color ID number returned was 2. which is blue. Table 6.3 shows the color index and names.

Table 6.3 Color Index and Names

Color Index	Color Names
0	Red
1	Green
2	Blue
3	Yellow
4	Magenta
5	Orange
6	White
7	Black
8	Pink
9	Gray
10	Light gray
11	Dark gray
12	Cyan

The HiTechnicColorSensor class defines several methods that return the color mode name and the color ID. There is also a SampleProvider class that stores the color ID value. To extract the color ID value from the sample, an array of floats is declared in line 17. It has the size of ColorVision.sample.size():

```
17      float X[] = new float[ColorVision.sampleSize()];
```

In line 19, the contents of the SampleProvider contain all the color ID values returned from the call to the ColorVision.getColorID() method

```
19      ColorVision.fetchSample(X,0);
```

The ColorVision.fetchSample(X,0) method accepts an array of floats and the offset. From the offset position, all the elements are copied to the array.

Lines 20 to 24 are used often. This loop reports all the values stored in the sample array:

```
20      for(int N = 0; N < ColorVision.sampleSize();N++)
21      {
22          Float Temp = new Float(X[N]);
23          Log.println("color sample value = " + Temp);
24      }
```

Digital Cameras Used to Detect and Track Color Objects

Color sensors are not the only device that can detect color. Digital cameras can be used to detect colors in a captured image. Considering that digital cameras can be small, they can be purchased as a component of a vision system for a robot or already embedded in the head of a robot.

Coupled with the digital cameras is an *image sensor.* This sensor is the device that converts the optical image into an electronic signal that serves as the "film" for the digital camera. They are sensitive to light and record the image. Each cell of the image is processed to gather all information to accurately re-create the image digitally. This image processing can also recognize colors of individual pixels. These devices can be used to detect color and perform object recognition (shape and color) to track that object. By using a camera, the object is detected from a stream of images. Detection means finding the colored object from one frame from a video stream. In this section, we discuss two color cameras: one embedded in Unit2 (RS Media Robosapien) and the Pixy vision sensor used by Unit1.

Tracking Colored Objects with RS Media

Tracking an object is the process of locating a moving or multiple objects over time using a camera. The idea is to detect the target objects in each consecutive video frame. Detection of the object can be difficult if the object(s) is moving fast relative to the camera. Some type of object recognition has to be performed to identify the targeting object in each frame. A number of object attributes can be used like size or shape, but the simplest one to use is color. The sensor has to be trained to recognize the object by letting it see the object, using the camera, so it can detect the color of the object. Once the target color of the object is detected in a frame, its location is accessed.

Unit2 is an RS Media Robosapien, a bipedal robot with a Linux kernel. It has several sensors and has an LCD screen. Figure 6.3 shows a diagram of the RS Media Robosapien, and Figure 6.4 shows its capability matrix in a LibreOffice spreadsheet. (LibreOffice is an open source office suite.)

BURT Translation in Listing 6.4 shows the pseudocode and Java fragment code translation for Unit2 to perform color object tracking. Listing 6.4 just contains the colorTrack() method.

Figure 6.3
Diagram of the RS Media
Robosapien

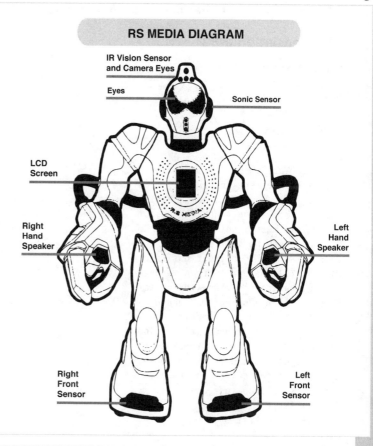

RS MEDIA DIAGRAM

IR Vision Sensor
and Camera Eyes

Eyes

Sonic Sensor

LCD
Screen

Right
Hand
Speaker

Left
Hand
Speaker

Right
Front
Sensor

Left
Front
Sensor

Figure 6.4
The capability matrix of Unit2 (RS Media
Robosapien) in a LibreOffice spreadsheet

Robot Name	Microcontroller	Speed	Memory
Unit2	ARM7	200 MHz	64 RAM
	16 Bit Processor		32 MB
Storage	**Communications**	**Sensors**	**Actuators**
SD Cards 1 GB	USB Port	3 Color/Light	11 W Speaker
		3 Sound Sensors	LCD Screen
		Microphone 20 kHz (chest)	
		Camera	
		3 Infrared detectors (6-24 in.)	
		2 Touch	
Motors	**End effectors**	**Mobility**	
12 – 11º DOF	Right Arm Gripper	Biped	
	Left Arm Gripper		

BURT Translation Listing 6.4   The `colorTrack()` Method

BURT Translation INPUT

```
Softbot   Frame
Name:   Unit2
Parts:
Sensor   Section:
Camera


Actions:
Step 1: Place in default position
Step 2: Put camera in tracking mode
Step 3: Tell camera to track a blue object
Step 4: Detect and track the object
Step 5: If the object was blue
Step 6: Wave the left arm
Step 7: Return to default position
Step 8: Tell camera to track a green object
Step 9: Detect and track the object
Step 10: If the object was green
Step 11: Wave the right arm
Step 12: Return to default position
Step 13: Tell camera to track a red object
Step 14: Detect and track the object
Step 15: If the object was red
Step 16: Continue to track
Step 17: Turn right 3 steps
Step 18: Turn left 3 steps
Step 19: Stop


Tasks:
Test the color sensor by tracking a colored object and then perform some movements.


End Frame
```

BURT Translations Output: Java Implementations

```java
1    private void colorTrack()
2    {
3        boolean BlueInColor = false;
4        boolean RedInColor = false;
5        boolean GreenInColor = false;
6        boolean BlackInColor = false;
7        int Blue = 0;
8        int Red = 2;
9        int Green = 1;
10       int Black = 4;
```

```
11
12          Unit2.restoreDefaultPosition();
13          Unit2.VISION.track();
14          BlueInColor = Unit2.VISION.getTarget(Blue);
15          if(BlueInColor){
16              Unit2.LEFT_ARM.wave();
17              Unit2.waitUntilStop();
18              Unit2.restoreDefaultPosition();
19          }
20          GreenInColor = Unit2.VISION.getTarget(Green);
21          if(GreenInColor){
22              Unit2.RIGHT_ARM.wave();
23              Unit2.waitUntilStop();
24              Unit2.restoreDefaultPosition();
25          }
26          RedInColor = Unit2.VISION.getTarget(Red);
27          if(RedInColor){
28              Unit2.VISION.track();
29              Unit2.waitUntilStop();
30              Unit2.VISION.track();
31              Unit2.waitUntilStop();
32              Unit2.WALK.turnRight(3);
33              Unit2.waitUntilStop();
34              Unit2.WALK.turnLeft(3);
35              Unit2.waitUntilStop();
36          }
37      }
```

For Unit2 to track an object, the object has to be placed within 2.5 cm of the camera and be stationary for Unit2 to learn the color first. Once the color is learned, Unit2 can detect any object of that color. Unit2 tracks the object's movements for 30 seconds and then it exits camera mode. Unit2 uses an infrared sensor to track the object within its range. It has a long and close range. To detect objects long range, it has to be within 24 inches from the sensor. At close range, an object has to be within 6 inches.

Unit2's vision is activated in line 13 with a call to the Unit2.VISION.track() method. It detects an object of a specified color and tracks that object with a call to getTarget(). It is called three times to detect and track three different colored objects:

```
14      BlueInColor = Unit2.VISION.getTarget(Blue);
        ...
20      GreenInColor = Unit2.VISION.getTarget(Green);
        ...
26      RedInColor = Unit2.VISION.getTarget(Red);
        ...
```

The getTarget() method returns a true or false if the target color object is detected and tracked. If true, Unit2 performs a few movements.

The method `Unit2.waitUntilStop();` is used several times. It is needed to make sure `Unit2` has completed the previous task before attempting another task. This is used often. A robot may take several compute cycles or even several minutes to perform a task. If the robot depends on the previous task's completion, using a wait is necessary. For example, if `Unit2` has to travel to a location before it is to detect an object at that location, depending on the distance, it may take the robot several minutes to arrive at that location. If the robot immediately attempts to detect a color object before it arrives at the location, it may detect the wrong object or no object at all. At that point, all bets are off, and the robot fails to complete its tasks.

Tracking Colored Objects with the Pixy Vision Sensor

As mentioned earlier, simple vision (to recognize an object by its color and track its location) for a robot is a system composed of several components and sensors:

- Recognition or color detections (color sensor)

- Distance or proximity detector (ultrasonic/infrared)

Use some image processing, and *Voila!* robot vision! RS Media has a nice vision system as we have discussed, but it comes up short in these ways:

- It's embedded in a robot.

- It can detect a single colored object.

- It does not provide information about the object.

The Pixy vision sensor is a digital color camera device that tracks objects by color and reports information about the object. Pixy has a dual-core dedicated processor that sends useful information like width, height, and x y location to the microcontroller at 50 frames per second for an image with 640×400 resolution. So information on detected objects is updated every 20 ms. The more frames per second or ms, the more precise documentation of the position of the object.

The Pixy can be connected to UART serial, I²C, USB bus interfaces; output can be digital/analog; and it can use Arduino and LEGO microcontrollers. Pixy can detect seven different color objects and track hundreds of items of the target colors at once. Using color codes, colors can be customized beyond the seven primary colors. Color codes combine colors to create unique objects to be detected. Table 6.4 lists the attributes of the Pixy and RS Media vision cameras.

Table 6.4   Some Attributes of the RS Media and Pixy Vision Cameras

Attribute	RS Media Vision Camera	Pixy Vision Sensor
# of colors	Red, blue, green, and skin tones	Indefinite
# of objects	1	Seven unique colored objects; indefinite objects real-time
Image sensor	N/A	OmniVision OV9715 Image sensor

Attribute	RS Media Vision Camera	Pixy Vision Sensor
Frame rate	N/A	50 fps
Frame resolution	N/A	1280 × 900, 640 × 400
Detection distance range	6 in. – 24 in.	Varying – 10 ft.
Training distance range	> 1 in.	6 in. – 20 in.
Dimensions of camera	Embedded	Pixy device: 53.34 × 50.5 × 35.56 mm
Weight of camera	Embedded	27 g
Microcontroller compatibility	Embedded	Arduino, Raspberry Pi, and BeagleBone Black
Refresh rate/tracking time	Tracking for 30 sec.	Update data 20 ms

Training Pixy to Detect Objects

Training Pixy to detect an object is similar to training the RS Media camera. The object has to be held so many inches in front of the camera so it can detect the color of the object. Once trained, the Pixy can detect other objects of the same color. With the Pixy there are two ways to do this:

 note

We used the Pixy for training but also used the PixyMon program for configuration.

- Use the PixyMon program.

- Use just the Pixy.

The Pixy has to be connected to the PC using a serial cable, and it must have a power supply. Run the PixyMon program to monitor what the camera sees. The Pixy sensor has an LED used to make sure the sensor recognizes the color and helps identify multiple objects to be detected. Once the button on the Pixy is pressed for 1 second, the LED sequences through several colors starting with white.

When the button is released, the camera is in light mode. The object should be held 6 inches to 20 inches in front of the lens in the center of the field of view. The LED should match the color of the object. For example, if the LED turns green when a green object is placed directly in front of Pixy, it has recognized that color. The LED flashes if successful and turns off if not successful. If successful, the button on the Pixy should be pressed and released; then the Pixy generates a statistical model of the colors in the object and stores them in flash. This statistical model is used to find objects of the same color.

If there are multiple objects to detect, each object must be associated with a signature, an identifier for the colors to be detected. Pixy's LED flashes through a sequence of seven colors; each is a signature:

- Red

- Orange

- Yellow

- Green

- Cyan

- Blue

- Violet

The first object is associated with Red (1), the second object is associated with Orange (2), and so on. The button is held until the desired color is flashed. PixyMon shows the video images of the sensor. As an object color is identified, the object is pixelated with that color (as shown in Figure 6.5).

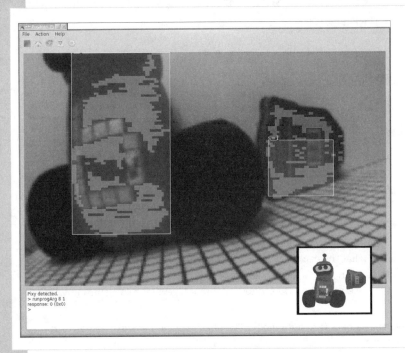

Figure 6.5
Screen capture of PixyMon where Pixy recognizes an object. Also shown is a picture of actual objects.

Programming the Pixy

Once the Pixy has been trained on the color objects it is to detect, it can be used to detect and track objects. Listing 6.5 is the BURT Translation that shows the pseudocode and Java code translation to perform color object tracking by the Pixy camera mounted on Unit1.

BURT Translation Listing 6.5   The Pixy to Perform Color Object Tracking

BURT Translation INPUT

```
Softbot   Frame
Name: Unit1
Parts:
 Sensor  Section:
Vision Sensor


Actions:
 Step 1: Initialize the sensor.
 Step 2: Detect and track objects.
 Step 3: If any objects were detected
 Step 4: Report the number of objects, width, height, xy location and color.


Tasks:
Test the vision sensor, track an object, and report basic information about an
object.


End Frame
```

BURT Translation Output: Java Implementations

```java
 1    //
 2    // begin license header
 3    //
 4    // This file is part of Pixy CMUcam5 or "Pixy" for short
 5    //
 6    // All Pixy source code is provided under the terms of the
 7    // GNU General Public License v2 (http://www.gnu.org/licenses/gpl-2.0.html).
 8    // Those wishing to use Pixy source code, software and/or
 9    // technologies under different licensing terms should contact us at
10    // cmucam@cs.cmu.edu. Such licensing terms are available for
11    // all portions of the Pixy codebase presented here.
12    //
13    // end license header
14    //
15    // This sketch is a good place to start if you're just getting started with
16    // Pixy and Arduino. This program simply prints the detected object blocks
17    // (including color codes) through the serial console. It uses the Arduino's
18    // ICSP port. For more information go here:
19    //
20    // http://cmucam.org/projects/cmucam5/wiki/Hooking_up_Pixy_
         to_a_Microcontroller_(like_an_Arduino)
21    //
22    // It prints the detected blocks once per second because printing all of the
23    // blocks for all 50 frames per second would overwhelm the Arduino's serial
         port.
24    //
```

```
25
26    #include <SPI.h>
27    #include <Pixy.h>
28
29    Pixy MyPixy;
30
31    void setup()
32    {
33        Serial.begin(9600);
34        Serial.print("Starting...\n");
35        MyPixy.init();
36    }
37
38    void loop()
39    {
40        static int I = 0;
41        int Index;
42        uint16_t Blocks;
43        char Buf[32];
44
45        Blocks = MyPixy.getBlocks();
46        if(Blocks)
47        {
48          I++;
49          if(I%50 == 0)
50          {
51              sprintf(Buf, "Detected %d:\n", Blocks);
52              Serial.print(Buf);
53              for(Index = 0; Index < Blocks; Index++)
54              {
55                  sprintf(Buf, "  Block %d: ", Index);
56                  Serial.print(Buf);
57                  MyPixy.Blocks[Index].print();
58                  if(MyPixy.Blocks[Index].signature == 1){
59                      Serial.print("color is red");
60                  }
61                  if(MyPixy.Blocks[Index].signature == 2){
62                  Serial.print("color is green");
63                  }
64                  if(MyPixy.Blocks[Index].signature == 3){
65                      Serial.print("color is blue");
66                  }
67              }
68          }
69        }
70    }
```

The Pixy sensor is declared in line 29 and is a Pixy object:

```
29    Pixy MyPixy;
```

The object MyPixy is initialized in the setup() function. The information about the detected objects is stored in the structure Blocks declared in Line 42 of type uint16_t.

In the loop() function, the getBlocks() method called in Line 45 returns the number of objects recognized in front of the video camera lens. If there are any recognized objects, the information for each object is reported. The Pixy is continuously detecting eight objects; the program is reporting the information for each object.

As mentioned earlier, Pixy processes 50 frames a second, and in each frame can be many objects that can be recognized. So information for every frame is not reported. Only when the I is divisible by 50, the number of Blocks is reported from a given frame.

The for loop cycles through the array block of Pixy objects recognized in the frame. In Line 57:

```
57  MyPixy.Blocks[Index].print();
```

The print() method reports the information for a particular Pixy object. The pieces of information stored for each object are:

- signature

- x

- y

- width

- height

The print() method reports this information. Each piece of information can also be reported using the attributes directly:

```
MyPixy.Blocks[Index].signature
MyPixy.Blocks[Index].x
MyPixy.Blocks[Index].y
MyPixy.Blocks[Index].width
MyPixy.Blocks[Index].height
```

Each attribute is of type uint16_t. By using signature, the color can also be reported. The Pixy was trained to detect three colors:

- signature 1: Red

- signature 2: Green

- signature 3: Blue

Depending on the signature, the correct color is reported:

```
58  if(MyPixy.Blocks[Index].signature == 1){
59     Serial.print(" color is red");
60  }
61  if(MyPixy.Blocks[Index].signature == 2){
```

```
62      Serial.print(" color is green");
63   }
64   if(MyPixy.Blocks[Index].signature == 3){
65      Serial.print(" color is blue");
66   }
```

A Closer Look at the Attributes

Figure 6.6 shows the field of view for the Pixy sensor.

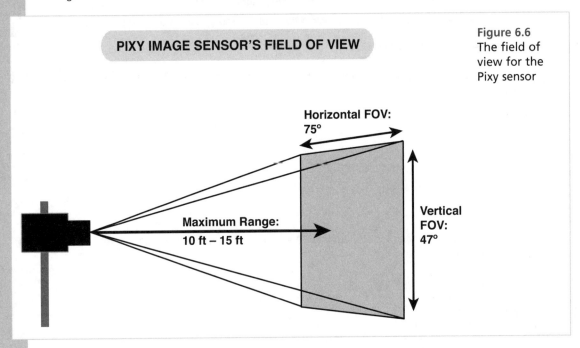

PIXY IMAGE SENSOR'S FIELD OF VIEW

Horizontal FOV:
75°

Maximum Range:
10 ft – 15 ft

Vertical
FOV:
47°

Figure 6.6
The field of
view for the
Pixy sensor

The objects detected are in the field of view of the camera. The horizontal FOV is 75° and the vertical FOV is 47°. With a 75° FOV, there is a maximum viewing distance of 10 feet. To extend the detection distance, a lens with a longer focal length can be used, but this would narrow the field of view. Pixy detects the xy location of the foci of the object. The minimum and maximum values for the width, height, x, and y attributes are listed in Table 6.5.

Table 6.5   Minimum and Maximum Values of Pixy Attributes

Max/Min	x	y	Height	Width
Minimum	0	0	1	1
Maximum	319	199	200	320

Ultrasonic Sensor

The ultrasonic sensor is a range finding sensor, an important component of a robot vision system. It measures the distance of an object from the actual sensor by generating sound waves that are not human audible, above 18 kHz. It then listens for the sound to be reflected back off objects in its sensing range. The time it takes to transmit and then receive the reflected wave tells us the time it takes for the wave to travel the distance. This time can then be converted to a distance measurement.

Many ultrasonic sensors are two in one; they transmit a signal and receive a signal. The transmitter sends high-frequency sound waves, and the receiver evaluates the echo received back. The sensor measures the time it takes to send the signal and receive the echo. This measurement is then converted to standard units (SI) such as inches, meters, or centimeters. The signal is really a sound pulse that stops when the sound returns. The width of the pulse is proportional to the distance the sound traveled, and the sound frequency range depends on the sensor. For example, industrial ultrasonic sensors operate between 25 kHz and 500 kHz.

The sensing frequency is inversely proportional to distance sensed. A 50 kHz sound wave may detect an object at 10 m or more; a 200 kHz sound wave is limited to detecting objects at about 1 m. So detecting objects farther away requires a lower frequency, and the higher frequencies are for objects closer to the sensor. Typical low-end ultrasonic sensors have frequencies of 30 kHz to 50 kHz. The ultrasonic sensors we use have a frequency of around 40 kHz and a sensing range of 2 to 3 meters.

Ultrasonic Sensor Limitations and Accuracy

The sensor emits a cone-shaped beam, and the length of the cone is the range of the sensor. It defines the FOV. If something is within the ultrasonic sensor's FOV, the sound waves are reflected off the object. But if the object is far away from the sensor, the sounds waves decrease the farther it travels, which affects the strength of the echo. This can mean that the receiver may not detect the echo or the reading may become unreliable. This is also true for objects at various extremes of the FOV. An ultrasonic sensor has a blind zone located nearest the face of the transmitter. This is an area where the sensor is unusable. This is the distance from the sensor where the beam hits a target and returns before the sound wave transmission is complete. The echo is not reliable. The outer edge of the blind zone is considered the minimum range an object can be from the transmitter. It is only a few centimeters.

What about objects to either side of the sensor? Will they be detected? It depends on the directivity of the sensor. The *directivity* is a measurement of the directional characteristics of the sound source generated by the transmitter. This measures how much sound is directed toward a specific area. Figure 6.7 (a) illustrates a typical sound energy radiation pattern's FOV at 50 kHz with sound decibels normalized on the x-axis. At various points in the pattern the distance and the level of decibels registered at that point can be determined. An object in the line of sight from the sound source registers the best performance at the maximum range. The lobes on each side show a shorter range.

 note

Objects measured above the maximum distance return a weak signal, which may also be unreliable.

Figure 6.7 (b) shows the FOV for an X ultrasonic sensor. Objects are located indicating its level of detection. As the distance increases the FOV also decreases to less than half at the maximum range.

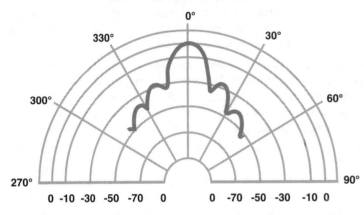

SOUND ENERGY RADIATION PATTERN'S FOV AND ULTRASONIC SENSOR'S FOV

Figure 6.7
A typical sound energy radiation pattern's FOV at 50 kHz; (b) FOV for an X ultrasonic sensor

(a) Sound energy radiation pattern's FOV at 50 kHz.

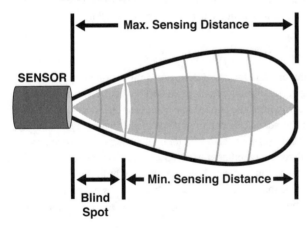

(b) FOV for a typical ultrasonic sensor.

When dealing with the limitations and accuracy of this sensor, directivity and ranges are not issues. Other limitations have to do with the material of the object and the angle the object is to the sensor and how that affects its accuracy. The accuracy of the sensor is how close the actual value is to the expected value. This calculation is in Chapter 5, Table 5.8 where A is the actual or measured value and E is the expected or true value. So an object at 60 cm distance from the ultrasonic sensor should produce a reading of 60 cm.

The accuracy of the EV3 ultrasonic sensor is documented at –/+1 cm. with a range of 3 cm to 250 cm. If the material surface absorbs sound, such as foam, cotton, rubber, this makes detecting more difficult compared to the object's surface made of more reflective materials such as plastic, steel, or glass. Sound absorbent materials can actually limit the maximum sensing distance. When using such objects, accuracy may be less than ideal or better by a few centimeters. The accuracy level should be tested for the application. The ultrasonic sensor performs best when the sensor sits high on the robot avoiding the sound waves hitting the floor and perpendicular to the object.

Figure 6.8 shows the various limitations of an ultrasonic sensor. The diagram shows the transmitter sending out a sound wave (or ping) that is reflected off an object (or wall). The first diagram (a) shows the sound wave returning an accurate reading. The object is directly in front of the sensor with the sensor parallel to the object and the beam hitting the object. The second diagram (b) illustrates *foreshortening* when the sensor is at an angle and the reading does not reflect an accurate distance. The pulse is wider (took a longer time to return because the distance has increased) than the original pulse. The third diagram (c) illustrates *specular reflection* when the wave hits the surface of the object or wall at an acute angle and then bounces away. The echo is not detected by the receiver at all. Odd-shaped objects can also reflect the sound waves in this way. The *crosstalk* is when multiple ultrasonic sensors in the system are in use simultaneously and all use the same frequency. The signals may be read by the wrong receiver as illustrated in (d). But this may be desirable where some sensors are in listening mode.

Figure 6.8
Various limitations of an ultrasonic sensor due to sound waves (a) accurate reading, (b) foreshortening, (c) specular reflection, (d) crosstalk

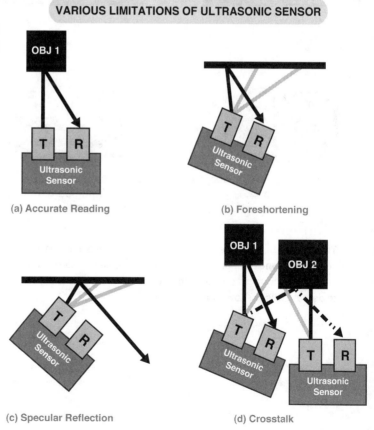

VARIOUS LIMITATIONS OF ULTRASONIC SENSOR

(a) Accurate Reading

(b) Foreshortening

(c) Specular Reflection

(d) Crosstalk

Regardless of the size of the object—a wide large object or a skinny one—with a wide cone and the way sound waves spread out, there would be no distinction between them.

Consider comparing an ultrasonic and an infrared long-range sensor that detects an object at 50 cm. The infrared sensor also has an emitter and a receiver, but it uses a single beam of IR light and triangulation to detect the distances of objects.

In Figure 6.9(b), using a narrow beam and a 10° FOV, an object can be located if the sensor is directly in its line of sight, 5° in either direction. Using this narrow beam it can detect doorways (no reading) where the ultrasonic sensor (Figure 6.9(b)) may detect the door frame. They can also be used to detect the width of objects. A robot equipped with both types of proximity sensors could better navigate paths through doorways and around a room with different sized obstacles. Some ultrasonic sensors can detect up to eight objects (returning multiple readings), but not all software avails the user of this option.

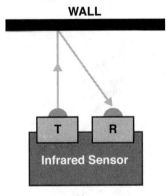

INFRARED AND ULTRASONIC SENSOR WAVES

WALL

WALL

T   R

T   R

Ultrasonic Sensor

Infrared Sensor

(a) Ultrasonic sensor reflects a diffused echo.

(b) Infrared sensor reflects a single beam utilizing triangulation.

Figure 6.9
An ultrasonic sensor detects a sound wave that is reflected off a target (a). An infrared sensor, a single beam utilizing triangulation is reflected off a target is detected.

The harder the object is to detect, whether its size, surface, or distance, the shorter the maximum sensing distance can be. Larger objects are easier to detect than smaller ones. Objects with smooth or polished surfaces reflect sound better than surfaces that are soft and porous, making them easier to detect. This is shown in Figure 6.10.

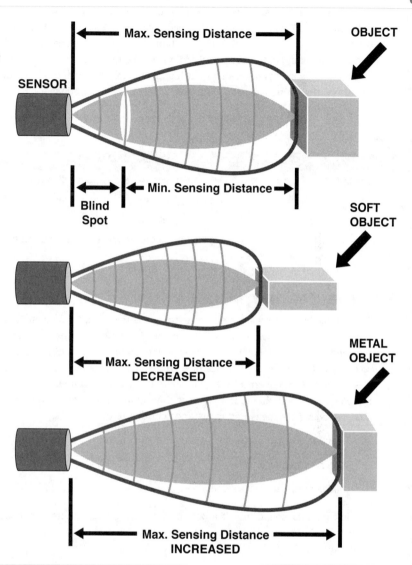

Figure 6.10
Various limitations of
an ultrasonic sensor due
to surface of the object

SENSOR

Max. Sensing Distance

OBJECT

Blind
Spot

Min. Sensing Distance

SOFT
OBJECT

Max. Sensing Distance
DECREASED

METAL
OBJECT

Max. Sensing Distance
INCREASED

Modes of the Ultrasonic Sensor

An ultrasonic sensor can be set to perform in different modes. The modes available depend on the manufacturer, the developer of libraries, and the software used to take advantage of its capabilities. An ultrasonic sensor can continuously send out sound waves or transmit a single ping and take a reading.

In continuous mode, the sound waves are sent out periodically at a regular interval. If an object is detected, the reading is stored. In single ping mode, the sensor sends out a single ping and takes a reading. If an object is detected, that reading is returned. Some sensors in this mode can detect multiple objects with a single ping where each reading is stored in a data structure like an array with the closet object reading in the first position. Active sensor mode for the ultrasonic sensor is the normal functioning of the sensor where it listens for the echo of the wave it transmitted. In passive mode, the sensor does not transmit. It just listens for echoes of other ultrasonic sensors. Table 6.6 lists the various ultrasonic sensor modes with a brief description.

Table 6.6 Ultrasonic Sensor Modes

Mode Type	Description
Continuous	Sound waves are sent out periodically at a regular interval.
Ping	A single ping is sent out and a reading is taken.
Active	Normal functioning.
Passive	No transmission; listens for echoes of other sensors.

Sample Readings

The sample for the ultrasonic sensor is the pulse width that represents time in the microseconds it took for the sound wave to reach the object and reflect back to the receiver. Once the reading is returned and stored it has to be converted into standard units (SI). We are interested in distances, so the standard units are distance measurements such as inches, feet, centimeters, and meters. It is necessary to convert microseconds into these standard units. As stated earlier, the width of the pulse is proportional to the distance the sound wave traveled to reach the object. The units for speed of sound are measured in distance per unit of time:

- 13503.9 inches per second

- 340 meters per second

- 34000 centimeters per second

In microseconds, sound travels at 29 microseconds per centimeter. So if our ultrasonic sensor returned 13340 microseconds, to determine how far the sound wave traveled, this calculation has to be performed:

13340 ms / 2 = 6670 ms

It is divided because the reading is a round trip and we want just the one-way distance.

6670 ms / 29 ms/cm = 230 cm = 2.3 m

So the object detected was 230 cm away, or 2.3 m away, from the sensor. This can be easily converted to inches if needed. So who performs this calculation? Some sensor libraries have functions that perform this calculation for you as a method for a built-in sensor or base sensor class. If not provided, the conversion function or calculation has to be written for the sensor class.

Data Types for Sensor Reading

 tip

An `int` should be used to store the microseconds or measurement reading returned. The microseconds can be a very large number. Once converted, the value should be stored as a `float` if decimal precision is required.

Calibration of the Ultrasonic Sensor

There are many different ways to calibrate an ultrasonic sensor to make sure the sensor is working properly and consistently produces the most accurate readings. Sensors should be tested to see how they perform in the system under development. Calibrating an ultrasonic sensor can be as simple as taking multiple readings of an object at the same distance to make sure the same readings are returned.

If the robot has multiple ultrasonic sensors, making one active and the other passive (crosstalk) allows each sensor to test the other. But the concern for accuracy also includes how the sensor's environment affects its performance. Some libraries have defined calibration methods for the ultrasonic sensor. If not, using continuous and ping modes in combination or separately will suffice.

For our birthday party scenario, consider the objects BR-1 may use its ultrasonic sensor to detect:

- Birthday cake

- Plates on a table

- Cups on a table

A birthday cake has a soft, porous surface. It may be difficult to detect the cake at farther ranges. The plates and cups (if they are made of glass or ceramic) have a reflective surface, but the plates are close to the table's surface. The good thing for the cake is there is no need to attempt to detect the cake at a long distance; it is on the table. So calibrating the ultrasonic sensor for this scenario requires that the sensor can detect these items at close range with varying surface types as shown in Figure 6.11.

Air temperature, humidity, and air pressure can also affect the speed of sound waves. If the ultrasonic sensor is to work in an environment with fluctuating temperature, for example, this may affect the readings. Molecules at higher temperatures have more energy and can vibrate faster, causing sound waves to travel more quickly. The speed of sound in room temperature air is 346 meters per second, much faster than 331 meters per second at freezing temperatures.

Accuracy concerning FOV, objects of different sizes, odd shapes, or surfaces requires different types of testing and calibration than for temperature and air pressure. A test plan should be created and measurements should be recorded and compared to expected values. Table 6.7 shows a sample testing of an ultrasonic sensor comparing the expected to the actual reading of the sensor on several objects at different locations.

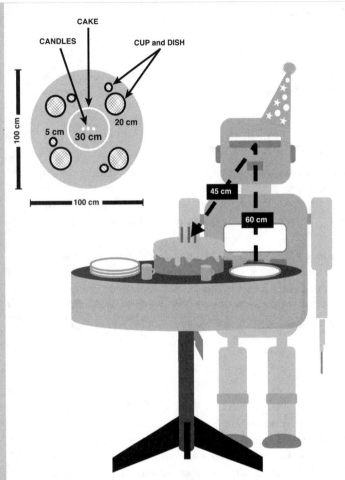

Figure 6.11
BR-1's situation to detect objects on the birthday table.

Table 6.7   Sample Testing of an Ultrasonic Sensor

Real Distance (cm)	1	2	5	10	15	20	25	30	40	50
1st Reading	6	7	6	12	17	21	27	30	40	50
2nd Reading	8	5	6	11	18	21	25	29	40	50
3rd Reading	4	4	6	10	17	21	25	29	41	50
4th Reading	7	5	6	11	18	21	25	29	42	50
5th Reading	5	4	6	11	19	21	25	30	40	49

Programming the Ultrasonic Sensor

In this section, we show BURT Translations for programming ultrasonic sensors for the EV3 and Arduino Uno microcontrollers. To program an ultrasonic sensor, the sensor must be initialized with the appropriate information and be prepared to receive input from its environment. The ultrasonic sensor may also require some type of calibration in the form of testing, validating, correcting, estimating the error, or determining the standards values for your system.

Listing 6.6 is the BURT Translation for programming the EV3 ultrasonic sensor. This BURT Translation shows the pseudocode for the actions of the ultrasonic sensor and the Java code translation. This is the main line to test the ultrasonic sensor to perform some basic operations. Listing 6.6 is a Softbot Frame for the BURT Translation for Unit1's ultrasonic sensor test and the Java code translation.

BURT Translation Listing 6.6   Softbot Frame for Unit1's Ultrasonic Sensor Test

BURT Translation INPUT

```
Softbot  Frame
Name:  Unit1
Parts:
Sensor  Section:
Ultrasonic  Sensor


Actions:
Step 1: Initialize and perform any necessary calibration to the
        ultrasonic sensor
Step 2: Test the ultrasonic sensor
Step 3: Report the distance and the modes


Tasks:
Test the ultrasonic sensor and perform some basic operations.
End Frame
```

BURT Translation Output: Java Implementations

```
49    public static void main(String [] args)  throws Exception
50    {
51        softbot SensorRobot = new softbot();
52        SensorRobot.testUltrasonicPing();
53        SensorRobot.getUSModes();
54        SensorRobot.closeLog();
55    }
```

In line 51, the softbot SensorRobot is declared. Three functions/methods are then called:

```
SensorRobot.testUltrasonicPing()
SensorRobot.getModes()
SensorRobot.closeLog()
```

In Listing 6.7 is the Java code for the softbot constructor.

Listing 6.7   The Softbot Constructor

```
8    public softbot() throws Exception
9    {
10       Log = new PrintWriter("Softbot.log");
11       Vision = new EV3UltrasonicSensor(SensorPort.S3);
12       Vision.enable();
13   }
```

Sensor initialization occurs in the constructor along with any necessary calibration. In line 11, like with all other sensors, the ultrasonic sensor is initialized with the sensor port and then enabled (line 12). No calibration was needed considering only a simple reading was to be performed.

BURT Translation in Listing 6.8 contains the "Softbot Frame" and Java translation for testUltrasonicPing().

BURT Translation Listing 6.8   The testUltrasonicPing() Method

BURT Translation INPUT

```
Softbot  Frame
Name:  Unit1
Parts:
Sensor  Section:
Ultrasonic  Sensor


Actions:
Step 1: Measure the distance to the object in front of the sensor
Step 2: Get the size of the sample
Step 3: Report the size of the sample to the log
Step 4: Report all sample information to the log


Tasks:
Ping the ultrasonic sensor and perform some basic operations.
End Frame
```

BURT Translations Output: Java Implementations

```
15    public SampleProvider testUltrasonicPing() throws Exception
16    {
17        UltrasonicSample = Vision.getDistanceMode();
18        float Samples[] = new float[UltrasonicSample.sampleSize()];
19        Log.println("sample size =" + UltrasonicSample.sampleSize());
20        UltrasonicSample.fetchSample(Samples,0);
21        for(int N = 0; N < UltrasonicSample.sampleSize();N++)
22        {
23            Float Temp = new Float(Samples[N]);
```

```
24              Log.println("ultrasonic value = " + Temp);
25          }
26          Log.println("exiting ultrasonic ping");
27          return UltrasonicSample;
28      {
```

Listing 6.8 shows the Java code for the testUltrasonicPing() method. Vision is an EV3UltrasonicSensor object. The sensor continuously sends out pulses. In line 17, the Vision.getDistanceMode() method listens for the echo and returns the measurement of the closest object to the sensor in meters. This value is stored in SampleProvider UltrasonicSample. In line 18, an array of floats (Samples) is declared of a size the number of samples. In line 20, the fetchSample() method assigns all elements in UltrasonicSample to Samples starting at offset 0. All values are then written to the log.

Listing 6.9 is the BURT Translation for the getUSModes() method.

BURT Translation Listing 6.9 The getUSModes() Method

BURT Translation INPUT

```
Softbot  Frame
Name:  Unit1
Parts:
Sensor  Section:
Ultrasonic  Sensor

Actions:
Step 1: Create an array that will hold all the possible mode names.
Step 2: Store the mode names in the array.
Step 3: Report the mode names to the log.

Tasks:
Report the mode names for the ultrasonic sensor.
End Frame.
```

BURT Translations Output: Java Implementations

```
31    public int getUSModes()
32    {
33        ArrayList<String> ModeNames;
34        ModeNames = new ArrayList<String>();
35        ModeNames = Vision.getAvailableModes();
36        Log.println("number of modes = " + ModeNames.size());
37        for(int N = 0; N < ModeNames.size();N++)
38        {
39            Log.println("ultrasonic mode = " + ModeNames.get(N));
40        }
41        return(1);
42    }
```

Here is the report generated by Unit1:

```
sample size = 1
ultrasonic value = 0.82500005
exiting ultrasonic ping
number of modes = 2
ultrasonic mode = Distance
ultrasonic mode = Listen
```

There was only one sample. The value of the sensor reading was 0.82500005 m. The EV3 ultrasonic sensor has two modes, Distance and Listen. In Distance mode, the sensor sends out a pulse, and listens for the echo to return the value. In Listen mode, the sensor does not send out a pulse or take a reading. It returns a value indicating other ultrasonic sensors are present.

Listing 6.10 contains the BURT Translation Input for taking readings of ultrasonic sensorsconnected to an Arduino Uno microcontroller.

BURT Translation Input Listing 6.10   Softbot Frame for Ultrasonic Sensor readings.

BURT Translation INPUT

```
Softbot  Frame
Name:  Unit1
Parts:
Sensor  Section:
Ultrasonic  Sensor

Actions:
Step 1: Set communication rate for the sensor
Step 2: Set the mode as Input or Output for the appropriate pins
Step 3: Send out a pulse
Step 4: Take a reading
Step 5: Convert the reading to centimeters
Step 6: Report the distance

Tasks:
Take a reading with the ultrasonic sensor and report the distance
End Frame
```

The next three programs are BURT Translation Outputs for three different ultrasonic sensors, each connected to an Arduino Uno microcontroller:

- HC-SR04

- Parallax Ping)))

- MaxBotix EZ1

Each ultrasonic sensor has important differences, which are discussed. An Arduino program (called a sketch) has two functions:

- `setup()`
- `loop()`

The `setup()` function is like a constructor. It is used to initialize variables, set pin modes, and so on. It only runs once after each time the Arduino board is powered up or reset. After the `setup()` function, the `loop()` loops indefinitely. So whatever is defined in the loop function is consecutively executed forever or until it is stopped. Listing 6.11 contains the BURT Translation C++ output for Ultrasonic Sensor 1, HC-SR04.

BURT Translation C++ Output Listing 6.11   C++ Translation for HC-SR04

BURT Translations Output: C++ Implementations

```
1    const int EchoPin = 3;
2    const int PingPin = 4;
3
4    void setup()
5    {
6        pinMode(PingPin,OUTPUT);
7        pinMode(EchoPin,INPUT);
8        Serial.begin(9600);
9    }
10
11   void loop()
12   {
13       long Cm;
14       Cm = ping();
15       Serial.println(Cm);
16       delay(100);
17   }
18
19   long ping()
20   {
21       long Time, Distance;
22       digitalWrite(PingPin, LOW);
23       delayMicroseconds(2);
24       digitalWrite(PingPin, HIGH);
25       delayMicroseconds(10);
26       digitalWrite(PingPin, LOW);
27       Time = pulseIn(EchoPin, HIGH);
28       Distance = microToCentimeters(Time);
29       return Distance;
30   }
31
32   long microToCentimeters(long Micro)
33   {
```

```
34        long Cm;
35        Cm = (Micro/2) / 29.1;
36        return Cm;
37    }
```

Figure 6.12 shows a diagram of this sensor.

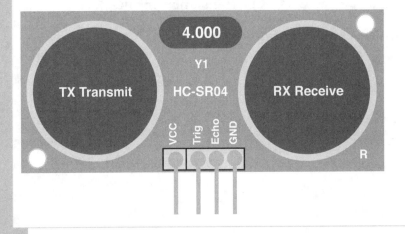

HC-SR04 ULTRASONIC SENSOR

Figure 6.12
HC-SR04 ultrasonic sensor

The HC-SR04 has four digital pins:

- **Vcc:** 5V Supply

- **Trig:** Trigger Pulse Input

- **Echo:** Echo Pulse Output

- **GND:** 0V Ground

The Vcc pin is connected to a 5V power source, and the GND is connected to the ground on the Arduino Uno board. We are interested in two pins:

- One used for pulse output. (PingPin is used to send out a pulse.)

- One used for pulse input. (EchoPin is used to read the pulse.)

The pinMode() function sets the specified digital pin to be used for either input or output. The pinMode() function accepts two parameters:

- pin

- mode

The `pin` is the number of the Arduino pin whose mode is to be set, and `mode` is the mode to be set. The `mode` is either `INPUT` or `OUTPUT`. A pin set to `INPUT` mode accepts input from whatever the pin is connected to (in this case an ultrasonic sensor), open to accept an electrical pulse. It is described as being in a *high-impedance state*. A pin set to `OUTPUT` mode sends out a pulse and is described as being in a low-impedance state. In line 6, the `PingPin` (pin 4) is set to be the `OUTPUT` pin, and in line 7 the `EchoPin` (pin 3) is set to be the `INPUT` pin. This means pin 4 is used to send out the ping to locate an object, and pin 3 is used to receive the reading if the ping is reflected off an object. `Serial.begin(9600)` sets the communication rate to 9600 in line 8.

The `loop()` function is defined in line 11 through 17. The `ping()` function is called in line 14 and the reading is returned. In line 15 the reading is reported. The `ping()` and `microToCentimeters()` functions are both user-defined functions. The `ping()` function sends out the pulse, takes a reading, and sends the reading (in microseconds) to the converted to centimeters. In `ping()function`:

- `digitalWrite()`

- `delayMicroseconds()`

function pairs are called twice. The `digitalWrite()` function writes a HIGH or LOW value to the specified digital pin. The `PingPin` has been configured as OUTPUT set with `pinMode()` in `setup()`; its voltage is set to LOW and then there is a 2 microsecond delay. The same `PingPin` is set to HIGH, meaning a 5V pulse is sent out and then there is a 10 microsecond delay. Then there is another LOW pulse write to the pin in line 26 to ensure a clean HIGH pulse. In line 27, `pulseIn()` listens for the HIGH pulse as an echo (reflected off an object) on `EchoPin` (pin 3) that was set for INPUT in the `setup()` function. The `pulseIn()` function waits for the pin to go **HIGH**, starts timing, and then waits for the pin to go **LOW** and then stops timing. It returns the length of the pulse in microseconds or 0 if no complete pulse is received in some specified time. This works on pulses from 10 microseconds to 3 minutes in length. Our pulse was 10 microseconds. If the `EchoPin` is already HIGH when the function is called (remember these readings are performed in a loop), it waits for the pin to go **LOW** and then **HIGH** before it starts counting again. Delays between readings are a good idea.

The duration of the pulse is returned, not the distance. So the microseconds reading has to be converted to a distance. The `microToCentimeters()` function converts the reading to centimeters. The microseconds reflect the length of the pulse from when it was sent out, bounced off some object in its path, and returned back to the receiver, `EchoPin`. To determine the distance, these microseconds are divided by 2. Then it is divided by 29.1, which is the number of microseconds it takes the sound to travel 1 centimeter. This value is returned. It is assigned to Cm in line 14 in the `loop()` function and then printed out to a serial output.

BURT Translation Output in Listing 6.12 contains the C++ translation for the Parallax Ping))) US2.

BURT Translation Output Listing 6.12   C++ Translation for Parallax Ping))) Ultrasonic Sensor

BURT Translations Output: C++ Implementations

```
1    const int PingPin = 5;
2
3    void setup()
4    {
```

```
 5        Serial.begin(9600);
 6    }
 7
 8    void loop()
 9    {
10        long Cm;
11        Cm = ping();
12        Serial.println(Cm);
13        delay(100);
14    }
15
16    long ping()
17    {
18        long Time, Distance;
19        pinMode(PingPin,OUTPUT);
20        digitalWrite(PingPin, LOW);
21        delayMicroseconds(2);
22        digitalWrite(PingPin, HIGH);
23        delayMicroseconds(5);
24        digitalWrite(PingPin, LOW);
25        pinMode(PingPin,INPUT);
26        Time = pulseIn(PingPin, HIGH);
27        Distance = microToCentimeters(Time);
28        return Distance;
29    }
30
31    long microToCentimeters(long Micro)
32    {
33        long Cm;
34        Cm = (Micro/2) / 29.1;
35        return Cm;
36    }
```

Figure 6.13 shows a diagram of the Parallax Ping))) ultrasonic sensor.

The Parallax Ping))) has three digital pins:

- **5V**: 5V Supply

- **SIG**: Signal I/O

- **GND**: 0V Ground

It only has one pin (SIG) to be used for INPUT or OUTPUT. So the program has to change a little to program this sensor. In the setup() function, only the communication speed is set:

```
3    void setup()
4    {
5        Serial.begin(9600);
6    }
```

Figure 6.13
Parallax Ping))) ultrasonic sensor

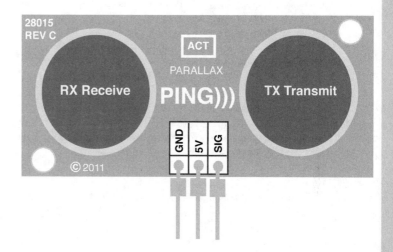

pinMode() is called in ping(). Since the same pin, PingPin (pin 5), has to be used for both input and output, the pin is reset to INPUT right before the reading is taken in line 25:

```
19      pinMode(PingPin,OUTPUT);
20      digitalWrite(PingPin, LOW);
21      delayMicroseconds(2);
22      digitalWrite(PingPin, HIGH);
23      delayMicroseconds(5);
24      digitalWrite(PingPin, LOW);
25      pinMode(PingPin,INPUT);
26      Time = pulseIn(PingPin, HIGH);
```

Everything else is identical.

Listing 6.13 shows the BURT Translation C++ Output for the MaxBotix EZ1 US3.

BURT Translation Output Listing 6.13   C++ Translation for MaxBotix EZ1 Ultrasonic Sensor

BURT Translation Output: C++ Implementations

```
1      const int PwPin = 7;
2
3      void setup()
4      {
5          Serial.begin(9600);
6      }
7
```

```
 8    void loop()
 9    {
10        long Cm;
11        Cm = ping();
12        Serial.println(Cm);
13        delay(100);
14    }
15
16    long ping()
17    {
18        long Time, Distance;
19        pinMode(PwPin,INPUT);
20        Time = pulseIn(PwPin, HIGH);
21        Distance = microToCentimeters(Time);
22        return Distance;
23    }
24
25    long microToCentimeters(long Micro)
26    {
27        long Cm;
28        Cm = Micro / 58;
29        return Cm;
30    }
```

Figure 6.14 shows a diagram of the MaxBotix EZ1 ultrasonic sensor.

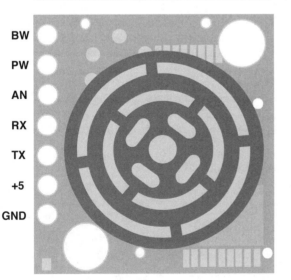

MAXBOTIX EZ1 ULTRASONIC SENSOR

BW
PW
AN
RX
TX
+5
GND

Figure 6.14
MaxBotix EZ1 ultrasonic sensor

The MaxBotix EZ1 has several pins including:

- **PW:** Pulse width output
- **AN:** Analog output
- **5V:** 2.5 to 5V supply
- **GND:** 0V Ground

This sensor supplies pins for PWM and analog outputs. PW is connected to an Arduino Uno digital pin, or the AN is connected to an analog pin for input. Like the EV3 ultrasonic sensor, EZ1 in free run mode, continuously pulses, so it does not need to be pinged. No `digitalWrite()` function needs to be called. In line 19, `pinMode()` puts `PwPin` in `INPUT` mode, and in line 20 the reading is taken. To convert the microseconds to centimeters, the reading is divided by 58. In this case, the pulse width is 58 microseconds per centimeter.

Compass Sensor Calculates Robot's Heading

Robots move around in their environment, going from one location to another performing tasks. Navigating their environment is crucial. Navigation is the robot's capability to determine its position in its environment and then to plan a path toward some goal location. As the robot goes from one location to another, it may be necessary to check to make sure it stays on its heading. Sometimes the robot may lose its direction due to slippage or the surface (terrain) of the environment, and a course correction becomes necessary by checking the current heading with the expected heading. A magnetic compass sensor measures the earth's magnetic (geomagnetic) field and calculates a heading angle. A magnetic compass works by detecting and orienting itself in the weak magnetic field on the surface of the earth thought to result from rotational forces of liquid iron in the earth's core. The sensor returns a value from 0 (true north) to 359°. The value represents the current heading angle of the robot as illustrated in Figure 6.15.

The sensor must be mounted horizontally on the robot. The compass should also be away from anything that emits electromagnetic fields such as motors, transformers, and inductors. Large conductive items, such as cars and fridges, can significantly alter magnetic fields affecting the reading. To minimize interference, the sensor should be placed at least 6 inches (15 cm) away from motors and at least 4 inches (10 cm) away from the microcontroller.

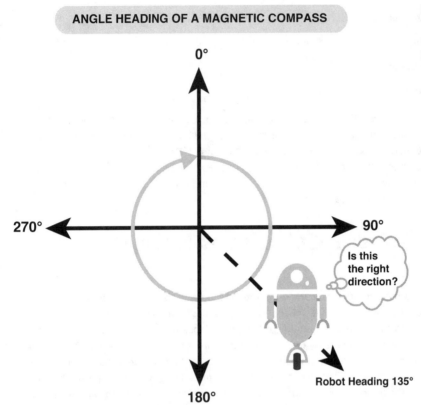

Figure 6.15
The angle heading of a magnetic compass

Programming the Compass

The HiTechnic magnetic compass uses the digital I²C communications protocol. A current heading is calculated to the nearest 1° and takes continuous readings 100 times per second. In this section, we show BURT Translations for programming this compass using the EV3 microcontroller. Listing 6.14 is the BURT Translation for programming the HiTechnic compass sensor using the `HiTechnicCompass` class. It contains the pseudocode for taking multiple compass readings and the Java code translation. This is the main line to test the compass sensor.

BURT Translation Listing 6.14   Unit1's Compass Sensor Test

BURT Translation INPUT

```
Softbot  Frame
Name:  Unit1
Parts:
Sensor  Section:
Compass Sensor
```

Actions:
Step 1: Initialize and perform any necessary calibration for the
        compass sensor
Step 2: Test the compass sensor
Step 3: Report the readings

Tasks:
Test the compass sensor by taking multiple readings.
End Frame

BURT Translations Output: Java Implementations

```
44    public static void main(String [] args)  throws Exception
45    {
46        softbot SensorRobot = new softbot();
47        SensorRobot.testCompass();
48        SensorRobot.closeLog();
49    }
```

Listing 6.15 shows the Java code for the constructor.

Listing 6.15   Java Code for Constructor

BURT Translations Output: Java Implementations

```
6     public softbot() throws Exception
7     {
8         Log = new PrintWriter("Softbot.log");
9         Compass = new HiTechnicCompass(SensorPort.S2);
10        calibrateCompass();
11    }
```

Listing 6.16 contains the BURT Translation for `calibrateCompass()`.

BURT Translation Listing 6.16   The `calibrateCompass()` Method

BURT Translation INPUT

```
Softbot  Frame
Name:  Unit1
Parts:
Sensor  Section:
Compass Sensor

Actions:
Step 1: Start calibrating the compass
Step 2: Report calibration has started
Step 3: Wait for 40 seconds
Step 4: Stop calibration
Step 5: Report calibration has stopped
```

Tasks:
Test the compass sensor by taking multiple readings.
End Frame

BURT Translations Output: Java Implementations

```
12    public void calibrateCompass() throws Exception
13    {
14        Compass.startCalibration();
15        Log.println("Starting calibration ...");
16        Thread.sleep(40000);
17        Compass.stopCalibration();
18        Log.println("Ending calibration ...");
19    }
```

The startCalibration() method in line 14 starts the calibration of the HiTechnic compass. To calibrate the compass, it has to be physically rotated 360 degrees twice. Each rotation should take 20 seconds. That is the reason for Thread.sleep(40000) in line 16.

Listing 6.17 contains the BURT Translation for the testCompass() method.

BURT Translation Listing 6.17   The testCompass() Method

BURT Translation INPUT

```
Softbot   Frame
Name:  Unit1
Parts:
Sensor   Section:
Compass Sensor

Actions:
Step 1: Get the compass sample
Step 2: Repeat 10 times
Step 3: Report the reading
Step 4: Wait awhile before getting another sample
Step 5: Get the compass sample
Step 6: End repeat

Tasks:
Test the compass sensor by taking multiple readings.
End Frame
```

BURT Translations Output: Java Implementations

```
22    public void testCompass() throws Exception
23    {
24        float X[] = new float[Compass.sampleSize()];
25        Compass.fetchSample(X,0);
26
27        for(int Count = 0; Count < 10;Count++)
28        {
29            Float Temp = new Float(X[0]);
30            Log.println("compass sample value = " + Temp);
31            Thread.sleep(5000);
32            Compass.fetchSample(X,0);
33        }
34    }
```

In line 25, the Compass.fetchSample() method retrieves the readings and places them in the array X. Only one value is returned and placed in the array. Since the compass takes 100 readings, a Thread.sleep(5000) method is used to pause between readings.

What's Ahead?

In this chapter, we talked about how to program different types of sensors. In Chapter 7,"Programming Motors and Servos," we will discuss the different types of motors, how to control gears, torque, and speed of motors and servos, and how to program the servos of a robot arm.

7

PROGRAMMING MOTORS AND SERVOS

Robot Sensitivity Training Lesson #7 *A robot's sensors can cause its grasp to exceed its reach.*

In Chapter 5, "A Close Look at Sensors," and Chapter 6, "Programming the Robot's Sensors," we discussed sensors: what they sense, how they sense, and of course how to program them. We saw that without sensors, a robot does not know the state of the environment, what is going on, or the current conditions. With that information, decisions are made and tasks can be performed for the situation. That activity is implemented by the actuators of the robot(s). Actuators are the mechanisms that allow robots to act upon the environment.

Actuators Are Output Transducers

In Chapter 5, we discuss how sensors, input transducers, can convert physical quantities into electrical energy. Figure 5.1 showed how a sound sensor transforms sound waves into an electrical signal measured in decibels. Figure 7.1 shows the conversion of an electromagnetic sound wave back to an acoustic wave performed by the output transducer, a loud speaker.

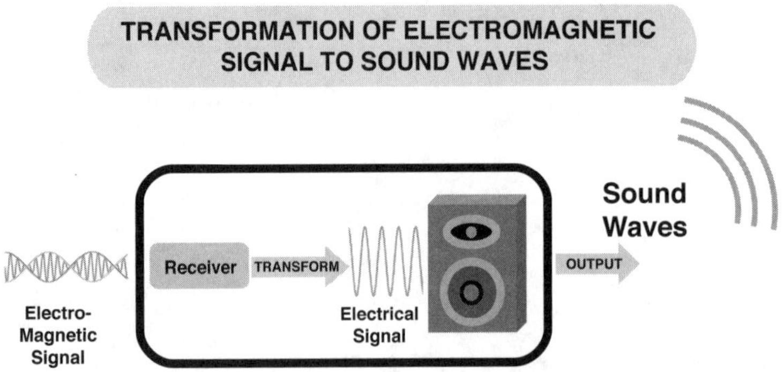

TRANSFORMATION OF ELECTROMAGNETIC
SIGNAL TO SOUND WAVES

Figure 7.1
A loud speaker converts an electromagnetic sound to an acoustic wave.

A special type of output transducer is an *actuator*. Actuators are devices that convert energy into physical motion such as:

- Potentiometer

- Accelerometer

- Linear and rotary motors

All these devices convert electrical signals to electromechanical energy. In this chapter we focus on rotary motors used as joints for robotic arms and end-effectors and to make robots mobile with wheels, tracks, and legs.

Different types of motors have different purposes. Some are more powerful than others and are good for moving large robots; others are better at more intricate motor skills, such as robot arms and end-effectors. When designing, building, and ultimately programming a robot, you should understand the different types of motors, their characteristics, and how they are used.

Motor Characteristics

Here are a few characteristics common to all motors. This is what should be considered when trying to determine what motors should be used for the robot components. Watch for these characteristics in the datasheets about a motor.

Voltage

The rated voltage of a motor is how much voltage is needed for the motor to operate at its peak efficiency. Most DC motors can operate a little above or below their rated voltage, but it's best not to plan on having it operate at those levels. The motor operating below the rated voltage reduces the motor's power; maybe other choices should be made. Operating above the rated voltage can burn out the motor. Motors should operate at their top speed/rated voltage at some point. Its slowest speed should be no more than 50% less than the rated voltage.

Current

Motors draw current depending on the load they're pulling. More load means more current usually. Every motor has a *stall current,* the current it draws when it's stopped by an opposing force. This current is much higher than the *running current,* or current that it draws under no load. The power supply for a motor should be capable of handling the stall current. When starting up, motors may draw near the stall current for a little while to overcome their inertia.

Speed

Motor speed is given in rotations per minute (RPMs).

Torque

Torque is the measure of a motor's capability to pull measured by the force a motor can pull when the opposing force (load) is attached to the motor's shaft. This measurement can be ft.-lb., lb-ft., oz.-in, in.-oz., g-cm (gram-centimeter), and any other weight to length variation.

Resistance

A motor can be rated in ohms. This is the resistance of the motor. Using Ohm's Law (voltage = current × resistance), you can calculate the motor's current draw.

Different Types of DC Motors

There are two types of actuators that motors can fall into:

- Linear

- Rotational

A linear actuator creates linear motion—that is, motion along one straight line. This type of actuator is important and useful for certain robot situations, but we do not discuss them in this book. Rotational actuators transform electrical energy into a rotating motion. Many types of motors are rotational actuators, and what distinguishes one type of motor from another are two main mechanical characteristics: torque and rotational speed. *Torque* is the force the motor can produce at a given distance and *rotational speed* has to do with how fast an object can rotate around an axis, the number of turns of the object divided by time. So this means a motor is defined by how much power it can exert and how fast it can go. The following three kinds of motor are discussed in this chapter:

- Direct current

- Servo

- Geared direct current

Although three motors are listed, all are different types of direct current (DC) motors. A servo motor is really a DC motor with a gear train.

Direct Current (DC) Motors

DC (direct current) motors and servos give robots physical action, such as walking if bipedal, rolling with tractor wheels, picking up and manipulating objects if the robots have arms. Considering those physical actions, when should a DC motor be used and when should a servo be used? What is the difference between a DC motor and a servo motor?

A DC motor comes in a variety of shapes and sizes, although most are cylindrical. Figure 7.2 shows a DC motor.

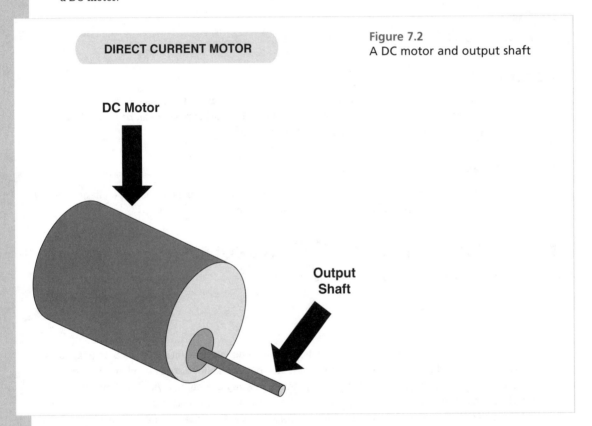

DIRECT CURRENT MOTOR

Figure 7.2
A DC motor and output shaft

DC Motor

Output Shaft

DC stands for "direct current," and that current causes the motor to rotate continuously at hundreds to tens of thousands RPM (rotations per minute). DC motors have two terminals (wires), and when direct current is applied to one terminal and grounding the other, the motor spins in one direction. When applying current to the other terminal and grounding the first terminal, the motor spins in the opposite direction. By switching the polarity of the terminals, the direction of the motor is reversed. Varying the current supplied to the motor varies the speed of the motor.

DC motors can be brushed and brushless. A brushed motor is made of an armature (a rotor), a commutator, brushes, an axle, and a field magnet. The armature or rotor is an electromagnet, and the field magnet is a permanent magnet. The commutator is a split-ring device wrapped around the axle that physically contacts the brushes, which are connected to opposite poles of the power source. Figure 7.3 shows the parts of the brushed DC motor and how it rotates.

Figure 7.3
Parts of a DC brushed motor and its rotation

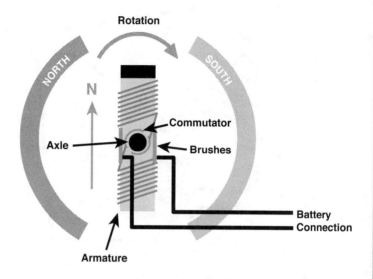

DC BRUSHED MOTOR PARTS AND ROTATION

The brushes charge the commutator inversely in polarity to the permanent magnet. This causes the armature to rotate. The rotation's direction can be clockwise and/or counterclockwise; it can be reversed by reversing the polarity of the brushes like reversing the leads on the battery. This process continues as long as power is supplied.

A brushless DC motor has four or more permanent magnets around the perimeter of the rotor in a cross pattern. The rotor bears the magnets, so it does not require connections, commutator, or brushes. Instead of these parts, the motor has control circuitry, like an encoder, to detect where the rotor is at certain times. Figure 7.4 shows a brushless DC motor.

For many applications and uses, brush and brushless motors are comparable. But each has pros and cons. Brushed motors are generally inexpensive and reliable. Brushless motors are pretty accurate for uses that have to do with positioning. Table 7.1 lists some pros and cons.

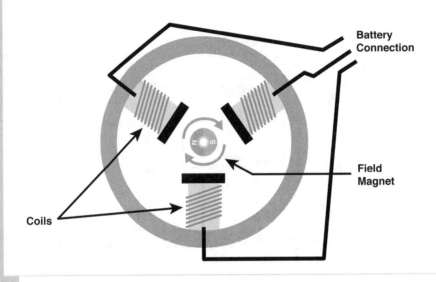

DC BRUSHLESS MOTOR PARTS AND ROTATION

Figure 7.4
Parts of a DC brushless motor and its rotation

Battery Connection

Field Magnet

Coils

Table 7.1 Pros and Cons of Brushed and Brushless DC Motors

Areas	Brushed Motors	Brushless Motors
Expense	**Pro:** Inexpensive.	**Con:** Higher cost of construction. Require a controller that can cost almost as the motor.
Reliability	**Pro:** Generally reliable in rough environments.	**Pro:** More reliable.
Accuracy	**Pro:** Good for high applications.	**Pro:** More accurate in positioning applications.
Operational issues	**Pro:** Somewhat extended operational life. Require few external components or no external components at all. **Con:** Inadequate heat dissipation caused by the rotor limitations.	**Pro:** Better heat dissipation. Low-noise (mechanical and electrical) operation. **Con:** Require control strategies that can be both complex and expensive.
Maintenance	**Con:** Require periodic maintenance; brushes must be cleaned and replaced for continued operation.	**Pro:** Require less and sometimes no maintenance.

Areas	Brushed Motors	Brushless Motors
Usage simplicity	**Pro:** Simple two-wire control; require fairly simple control or no control at all in fixed-speed designs.	**Con:** More difficult to control.
Torque/ speed	**Con:** As speed increases, brush friction increases and viable torque decreases. Low speed range due to limitations imposed by the brushes.	**Pro:** Capability to maintain or increase torque at various speeds. Higher speed ranges.
Power	**Con:** Power consumption problems.	**Pro:** No power loss across brushes. Components significantly more efficient. High output power. **Con:** Too much heat weakens the magnets and may cause damage.
Size	**Pro:** Varying sizes.	**Pro:** Small size.

Speed and Torque

Motors have to be controlled for them to be useful for a robot. They have to start, stop, and increase and decrease speed. The speed of the DC motor can be easily controlled by controlling the power level or the voltage supplied to the motor. The higher the voltage, the higher speed the motor tries to go. Pulse Width Modulation (PWM) is a way to control the voltage and therefore the speed of the motor. With the basic PWM, the operating power to the motors is turned on and off to modulate the current going to the motor. The ratio of the "on" time to the "off" time determines the speed of the motor. This is the *duty cycle*, the percentage the motor is on to the percentage the motor is off. Switching the power on and off fast enough makes it seem that the motor is slowing down without stuttering. When using this approach, not only is there a reduction of speed, there is also a proportional decrease in the torque output. There is an inverse relationship between torque and speed: As speed increases torque is reduced; as torque increases speed reduces. Figure 7.5 shows the relationship between torque, speed, and velocity.

Torque is the angular force that produces motor rotation measured in

- Pounds per feet (lb/in)

- Ounces per inch (oz/in)

- Newtons per meter (N/m)

RELATIONSHIP BETWEEN TORQUE, SPEED, AND VELOCITY

Figure 7.5
The relationship between torque, speed, and velocity

Torque is not a constant value; it can have different values depending on the given information or condition. For example, the stall torque is the measurement of torque when it is at its maximum. That means the greatest torque has been used to make the motor rotate from a standstill state. A *full load* is the amount of torque needed to produce the rated power (horsepower) at the full speed of the motor. The stall torque is often higher than the full load torque. Sometimes these values are supplied by the manufacturer on the motor's data sheet. Sometimes other information is supplied and then stall and full load torque values can be calculated:

Full Load Torque = (Horse_Power * 5252) / RPM

Horsepower is measured in watts. To calculate stall torque:

Stall Torque = $Power_{max}$ / RPM

No load speed is the measurement of the motor at its highest speed rotating freely with no torque. Between the stall and no load is the rated or *nominal* torque, which is the maximum torque that ensures continuous operation without problems. This is approximately about half of stall. The *startup torque* is the amount required by a robot to perform a task and should be around 20% to 30% of the maximum torque of the motor. Figure 7.6 shows the relationship of stall torque and no load falls on the graph.

Figure 7.6
The relationship between stall, nominal, and no load torque

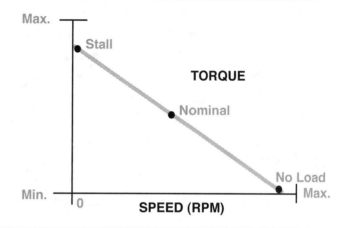

RELATIONSHIP BETWEEN STALL, NOMINAL, AND NO LOAD TORQUE

Motors with Gears

Another way to reduce the speed of the motor without reducing the voltage and therefore the torque is by using gears. A motor rotates with high speed, but the torque generated by its rotation is not enough to move even a light load. The gearbox or gear mechanism takes high input speed of the motor and makes the output speed slower than the original input speed. This increases the torque. The rotation speed can be increased or decreased along with the corresponding increase or decrease in the torque. A number of things can be accomplished when transmitting power using gears:

- Change the rotational direction

- Change the angle of rotation

- Convert rotational motion into linear motion

- Change the location of rotational motion

Gear Fundamentals

Gearing down means using gears to slow down the output of the motor before it is applied to a shaft. Consider Figure 7.7 (a). The small gear (X) is called the pinion; it has 16 teeth. The large gear (Y), called the wheel, has 32 teeth. That's double the teeth on the pinion. The *ratio of the gears* is the number of teeth on the wheel divided by the number of teeth on the pinion. So, there is a ratio of 2 to 1 between X and Y. This means that for each revolution of the pinion, the wheel makes a half turn. If a motor is connected to the pinion and the output shaft to the wheel, the motor speed is cut in half at the output shaft. Figure 7.7 (b) shows this configuration on the robot arm, which doubled the torque output.

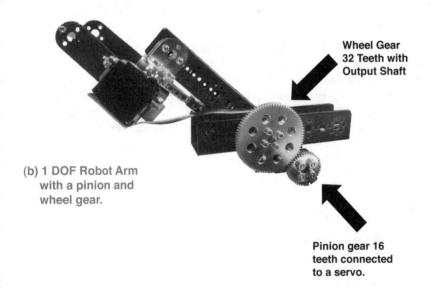

PINION AND WHEEL GEARS ON A ROBOT ARM

Full Turn Clockwise

Half Turn Counter-
Clockwise

Pinion Gear
16 Teeth

X

Y

Wheel Gear
32 Teeth

(a) Pinion gear (X) and
wheel gear (Y) has a
2 to 1 ratio.

Figure 7.7
Pinion and wheel
gears: (a) pinion (X) has
16 teeth and wheel (Y)
has 32 teeth; (b) photo
of a robot arm using a
pinion and wheel

Wheel Gear
32 Teeth with
Output Shaft

(b) 1 DOF Robot Arm
with a pinion and
wheel gear.

Pinion gear 16
teeth connected
to a servo.

The speed and torque of the wheel can be calculated given the speed or torque of the pinion:

$$T_W = e(T_P R)$$
$$S_W = e(S_P/R)$$

where T_W and S_W are the torque and speed of the wheel. T_P and S_P are the torque and speed of the pinion, R is the gear ratio, and e is the gearbox efficiency, which is a constant between 0 and 1.

For example, a pair of gears:

$R = 2$
$T_P = 10$ inch-pound
$e = .8$ (which is at 80% efficiency),
$S_P = 200$ RMP

The output wheel torque will be:

16 pounds per inch $= .8\ (10 \times 2)$

The speed of the wheel will be:

80 RPM $= .8\ (200/2)$

If a higher gear reduction is needed, adding more gears won't do it and the efficiency will drop. The reduction of any gear is a function of the gear ratio, the number of teeth on the output gear connected to the shaft (follower gear Y) divided by the number of teeth on the input gear connected to the motor (driver gear X). Adding gears between the driver and the follower, called *idlers*, does not cause an increase in the gear ratio between the same two gears. To have a larger gear ratio, gears can be layered. Attach a pinion (smaller gear X_2) to the wheel shaft Y and use that second pinion to drive the larger wheel (Y_2) as shown in Figure 7.8.

Figure 7.8
Pinion and wheel gears layering

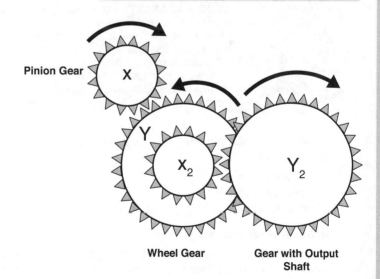

LAYERING WITH PINION AND WHEEL GEARS

Pinion Gear — X

Y — X_2 — Y_2

Wheel Gear   Gear with Output Shaft

As far as size, say:

$X = X_2$
$Y = Y_2$

The gear ratio is shown. X_2 attached to Y rotates at the same speed. Therefore, X revolves 4 times (input rotations) to every 1 revolution of Y (output rotation), 4:1 reduction. This is commonly done inside gearboxes. The gear efficiency should be supplied by the manufacturer of the gear on the data sheet based on the type of gear. Table 7.2 is a list of typical gears and their estimated efficiency constants. The gear ratio or reduction should also be supplied by the manufacturer.

Table 7.2 Typical Gears and Estimated Efficiency Constants

Gear Type	Description	# of Teeth/ Threads	Estimated Gear Efficiency
	SPUR: Radial teeth (teeth that spread from the center outward) parallel to the axis; commonly used simple gear; gears mesh together if fitted to parallel shafts.	8, 16, 24, 40	~90% Highest Possible Efficiency
	BEVEL: Shaped like a right circular cone with most of its tip cut off; angle between the shafts can be anything except zero or 180 degrees. Bevel gears with equal numbers of teeth and shaft axes at 90 degrees.	12, 20, 36	~70 % Low Efficiency
	WORM & GEAR SET: A worm is a gear with one or more cylindrical, screw-like threads and a face width that is usually greater than its diameter. Worm gears differ from spur gears in that their teeth are somewhat different in shape to mate with worms. Worms have threads where gears have teeth.	teeth/threads ratio teeth/1 thread	Wide range efficiency 50% – 90%

Calculating the total gear reduction of a gearbox or drive train is simple. Multiply all the gear ratios of the gear sets. For example:

Gear set #1: 4:1

Gear set #2: 3:1

Gear set #3: 6:1

with a total gear reduction: 4 * 3 * 6 = 72:1

So if a motor has a speed of 3000 RPM with a gearbox of 72:1, then the RPM for the motor is 3000/72 = 42 RPM. Use the gear ratio to determine the effect on the torque of the motor.

Gears can also change the rotational direction. Two gears, like the wheel and pinion, can reverse the rotational output. So if the input wheel is turning counterclockwise, then output gear turns clockwise. But to have the output wheel to turn in the same direction, add an idler gear between them as in Figure 7.9.

Figure 7.9
Input (driver) wheel turning counterclockwise causes the output wheel to turn clockwise. Adding an idler wheel between them causes both wheels to turn in the same direction.

DRIVERS, IDLERS, AND FOLLOWER GEARS

OPPOSITE DIRECTIONS

Driver   Follower

SAME DIRECTION

Driver   Idler   Follower

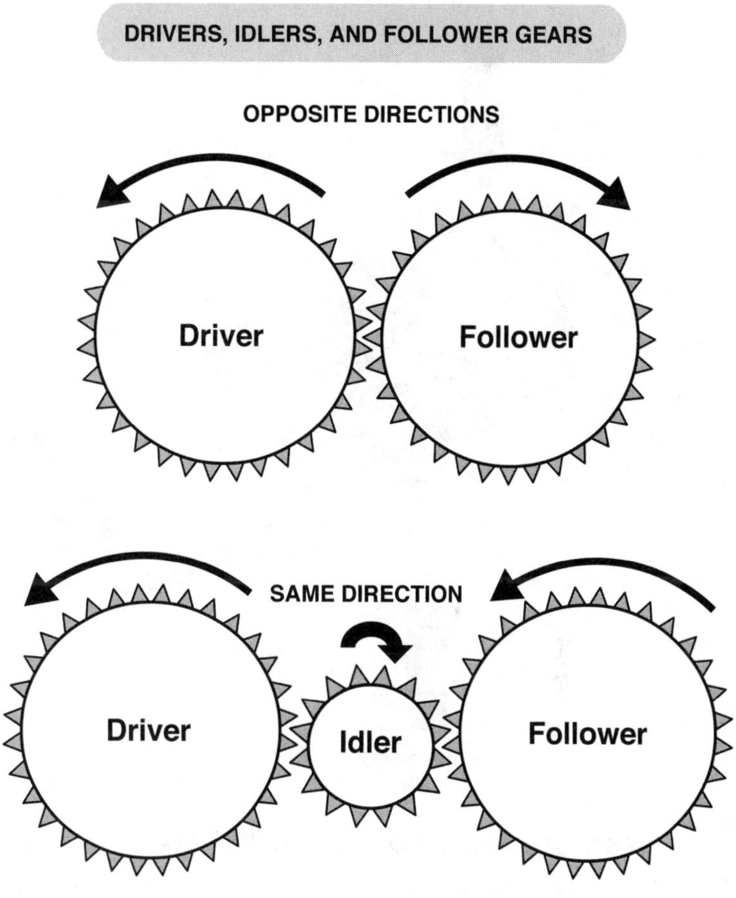

The output wheel now turns counterclockwise, the same as the input gear. But what if you have a chain of say four or five gears touching? The rule is this: With an odd number of gears, rotation (input and output gears) is in the same direction, and with an even number of gears, they are counterrotational and the gear ratio stays the same. The input and output gears are accessed; all idler gears are ignored. The gear efficiency is:

$$\text{Total\_gear\_efficiency} = \text{gear\_type\_efficiency}^{\text{\# of gears} -1}$$

DC Motor with Gearhead

Some DC motors have a gearbox and are called *gearhead* DC motors. Figure 7.10 shows the DC motor with a gearbox. The gearbox is on the shaft end on the motor. The gearbox makes the output of the motor or the turning of the shaft slower and more powerful without lowering the voltage.

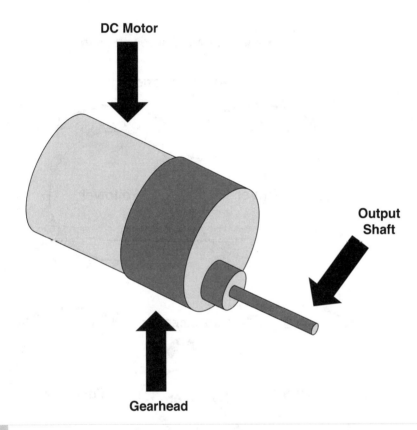

Figure 7.10
The DC motor with
gearbox on shaft end

DIRECT CURRENT MOTOR WITH A GEARHEAD

DC Motor

**Output
Shaft**

Gearhead

Servos: DC Motors with Geartrain

Servo motors are a variation of the gearhead motor. They are coupled with a potentiometer (pot). A pot gives feedback on the motor's position. Servo motors are used for precise positioning and speed control at high torques. They consist of a motor, position sensor, and controller. Servos have three wires. The first two are for power and ground, and the third is a digital control line. The digital control line is used to set the position of a servo. The small DC motor sensor, and controller are all inside a single plastic housing. Servo motors are available in power ratings from fractions of a watt up to a few 100 watts. A servo motor's output shaft can be moved to a specific angular position or rotation speed by sending it a coded signal. The servo motor maintains the position of the shaft or speed as long as the same coded signal is applied. When the coded signal changes, the shaft changes too.

A servo motor turns 90° in either direction, with a maximum rotation of 180°. There are servos with a 360° rotation. A normal servo motor cannot rotate any more than that due to a built-in mechanical stop. Three wires are taken out of a servo: positive, ground, and control wire. A servo motor is controlled by sending a Pulse Width Modulated (PWM) signal through the control wire. A pulse is sent every 20 ms. The duration of the pulses determines the position of the shaft. For example, a pulse of 1 ms, minimum pulse, moves to 0°; a pulse of 2 ms, its maximum pulse, moves the shaft to 180°; and a pulse of 1.5 ms, neutral position, moves the shaft to 90°. This is shown in Figure 7.11.

Figure 7.11
The pulse duration and the corresponding position of the shaft

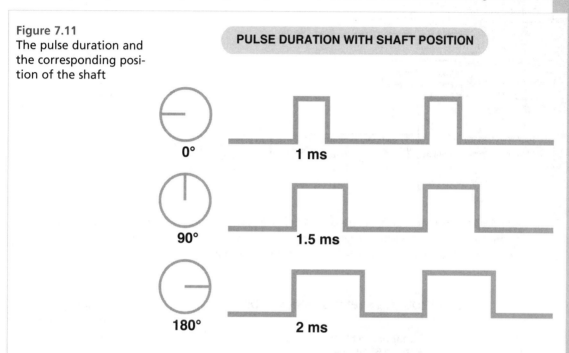

PULSE DURATION WITH SHAFT POSITION

0° — 1 ms

90° — 1.5 ms

180° — 2 ms

If an external force is trying to change the position of the shaft, the motor resists the change. Pulses need to be repeated for the motor to hold the position. The potentiometer measures the position of the output shaft at all times so the controller can correctly place and maintain the servo shaft at the desired position.

The servo uses a closed-loop control, which gives feedback to the controller so adjustments can be made to maintain the desired position or speed of the motor's output shaft. When the pulse changes, an error is calculated between the current position/speed and the new position/speed. The electrical pulse is first fed to a voltage converter. The current rotational position or speed of the output shaft is read by the tachometer/pot that produces a voltage. The tachometer/pot serves as an electromechanical device that converts mechanical rotation into an electrical pulse. The rotational position is related to the absolute angle of the output shaft. The signal representing the current position/speed and the signal representing the new or commanded position/speed are both fed to an Error Amplifier (EA). The output of the EA drives the motor. The EA determines the difference between the voltages: If the output is negative, the motor is moved in one direction, and if the output is positive, the motor is moved in the other direction. The greater the difference, the greater the voltage and the more the motor moves or turns, speeds up, or slows down. When the motor turns, this engages the gear system that turns the output shaft to the commanded position. The servo motor continuously makes adjustments, resisting any movements to maintain that feedback operation that occurs every so many milliseconds. This process is depicted in Figure 7.12.

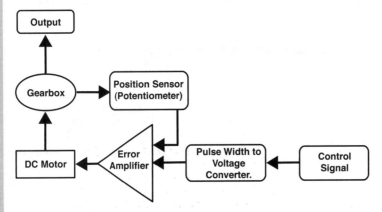

SERVOMECHANISM'S CLOSED LOOP CONTROL

Figure 7.12
The closed-loop control for a servomechanism

The closed-loop system monitors the error difference between how fast/slow the robot's motor is going, how fast/slow it should be going, and makes adjustments to the motor's power level if necessary. Figure 7.13 shows an example of controlling the speed of the servo.

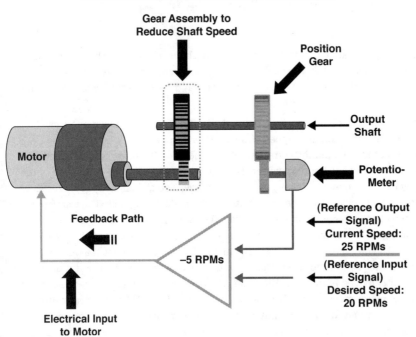

CONTROLLING SPEED OF THE MOTOR

Figure 7.13
An example of controlling the speed of the servo.

What Are Encoders For?

Encoders are used to track the turning of motor shafts to generate the digital position of the shaft and other motion information. They convert motion into a sequence of digital pulses. Encoders can be built in to the motors like servos or they can be external as in the case of the DC motor where they are mounted on the motor's shaft. Figure 7.14 shows the location of the encoder for the DC motor and a LEGO servo.

Figure 7.14
Encoders for (a) DC motor; (b) inside a LEGO servo

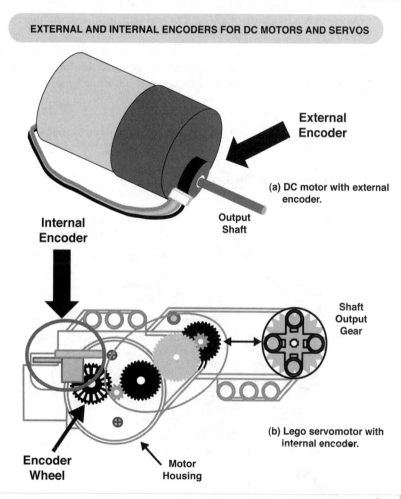

EXTERNAL AND INTERNAL ENCODERS FOR DC MOTORS AND SERVOS

External Encoder

(a) DC motor with external encoder.

Output Shaft

Internal Encoder

Shaft Output Gear

(b) Lego servomotor with internal encoder.

Encoder Wheel

Motor Housing

Encoders can have a linear or rotary configuration, but the most common is the rotary type. There are two basic forms of rotary encoders: incremental and absolute. Incremental encoders produce a digital pulse as the shaft rotates, allowing the measurement of the relative position of the shaft. Absolute encoders have a unique digital word for each rotational position of the shaft. These are often paired with permanent-magnet brushless motors for servos.

Most of the rotary encoders are made of a glass or plastic code disk. Some have a photographically deposited radial pattern organized in tracks. The NXT LEGO servo has a black wheel with 12 openings or slits (refer to Figure 7.14 [b]). As the motor spins, the encoder wheel also spins interrupting a beam of light on one side of the wheel that is detected by sensors on the other side of the wheel. The rotation generates a series of on and off states. Bits can be combined to produce a single byte output of *n* bits as a parallel transmission.

The Tetrix encoders (manufactured by Pitsco) for DC motors are quadrature optical incremental encoders. This means the encoders output a relative position and the direction of the motor. For this type of encoder, the beam of light passes through an encoder that splits the light to produce a second beam of light 90° out of phase. Light passes from A and B channels through the wheel onto the photodiode array. The wheel rotation creates a light-dark pattern through the clear and opaque wheel segments. The pattern is read and processed by a photodiode array and decoding circuitry. Beams A and B are each received by a separate diode and converted into two signals 90° out of phase with respect to the other sensor. This output is called quadrature output. It's then fed to a processor that can process the signal to determine the number of pulses, direction, speed, and other information. Figure 7.15 shows the incremental encoder disk, light, and sensors.

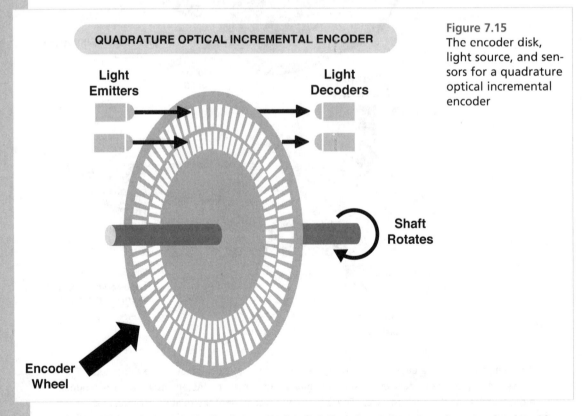

Figure 7.15
The encoder disk, light source, and sensors for a quadrature optical incremental encoder

The LEGO servo encoder wheel only has one level of slits where other servos have two levels, with a sensor for each level.

Motor Configurations: Direct and Indirect Drivetrains

There are different ways to connect the motors to the components that move or rotate. The way they are connected determines how well the motor mechanism performs in the situation. The drivetrain is a mechanical system that transfers rotational power from the motor to the drive wheels. There are two types of drivetrains:

- Direct transfer

- Indirect transfer

For a direct drive transfer, the motor shaft is directly connected to the drive wheel. Direct drive transfer means there is no speed torque changes between the motor and the wheel and is mechanically efficient. But the parts in the direct drive system can wear down and break.

For an indirect drive transfer system, the motor shaft is connected to the drive wheel shaft through a gear train and maybe belts or chains. The gear train and belt reduce the speed of the motor while proportionately increasing motor torque. This also acts as a shock absorber when the drive wheels are stuck or under heavy load. Figure 7.16 shows the two types of drivetrains.

Figure 7.16
Direct and indirect drivetrains.

DIRECT AND INDIRECT DRIVETRAINS

Wheel

Encoder

(a) Direct drivetrain where shaft is directly connected to wheel.

Wheel

Output Shaft

Gears

(b) Indirect drivetrain where shaft is indirectly connected to wheel using gears and an output shaft.

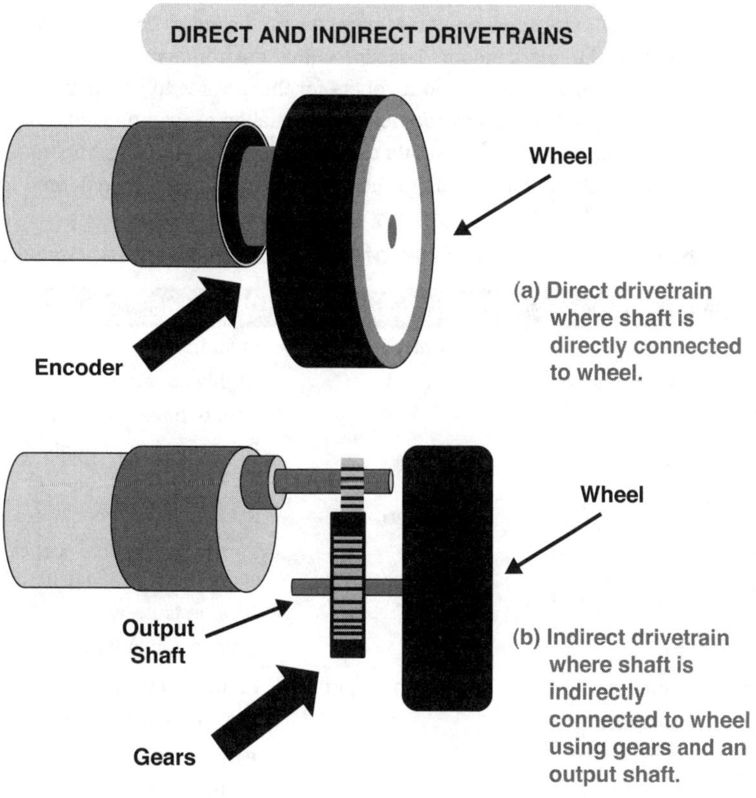

The drivetrains play an important role in the robot's mobility and joints in the robot arm. Table 7.3 describes the drivetrains, their advantages, and disadvantages.

Table 7.3 Lists of Drivetrains and Their Advantages

DRIVETRAINS	DESCRIPTION	ADVANTAGES	DISADVANTAGES
Direct	The motor shaft is directly connected to the drive wheel.	There are no speed or torque changes between the motor and the wheel.	Parts can wear down and break.
Indirect	The motor shaft is connected to the drive wheel shaft through a gear train and maybe belts or chains.	It reduces the speed of the motor while proportionately increasing motor torque; it acts as a shock absorber when the drive wheels are stuck or under heavy load.	The overall speed is reduced; a small gear can drive a larger gear to increase speed and decrease torque but may damage the DC motor's internal gearbox.

Terrain Challenge for Indoor and Outdoor Robots

Some of the robots discussed in this book are meant for indoor scenarios and situations. The birthday party scenario and Midamba's predicament both take place indoors. Many robots are programmed to work indoors because indoor environments are safer and more predictable. The premise of programming autonomous robots for this book is to program them for scenarios totally under control and predictable environments. The challenges can be well thought out, and success is probable. But even indoor environments can introduce challenges for robots. Table 7.4 lists the different types of environments and some of the factors that should be considered for motors.

Table 7.4 Different Types of Robot Environments and Some Important Factors

Robot Environment	Challenges	Description
Indoor Most common	Controlled	Include mazes.
		Highly constrained, few surprises.
		Robots have a specific task.
		Drivetrains are simple using wheels directly connected to motors.
	Uncontrolled	Robots roam unconstrained or under moderate constraint.
		Terrain issues like carpet, raised door thresholds, carpet-to-floor, floor-to-carpet transitions, stairs.
		Ground clutter.
		Obstacle avoidance and navigation.
Outdoor Complicates robot design	Weatherizing	Protection from:
		▪ Temperature extreme
		▪ Rain/moisture
		▪ Humidity

Robot Environment	Challenges	Description
	Protection from dirt	Dirt stays away from circuitry, gears.
	Vibration damping	Lessen the effect of vibration on mounted motors, circuit boards, so they do not come loose.

Dealing with Terrain Challenges

The challenges of terrain can be addressed by steering and driving systems, the type of mobility, as well as the torque and speed of the robot. Some terrains are better negotiated with legs, especially if climbing is needed (like stairs). Wheels and tracks are better with other types of terrain. If an object is to be transported, which type of mobility is better for the task? A robot with legs, wheels, or tractor wheels?

The size and weight of the robot becomes a factor in making these choices. Smaller robots may have fewer problems in controlled environments because they are faster and can easily get around known small obstacles. But on the other hand, if a robot has to travel from waypoint A from waypoint C in a certain time frame and over a thick carpeted area to reach waypoint C, would a larger robot be better? A smaller robot's batteries may become depleted trying to supply its motors a high constant current to maintain a speed at the cost of torque. Those wheels are rotating, but the robot is not going very far. Maybe a larger robot with bigger wheels and more powerful motors is a better fit. It is less affected by the terrain, and the larger wheels cause the robot to travel farther but rotate slower. But bigger, heavier robots also have challenges. Larger robots draw more current. Farther distances mean drawing more current from the batteries, which are already being taxed. When the motor stalls, it draws a large amount of current. Does the larger robot have indirect drive that can help with lessening the load on the motors?

What type of mobility? If wheels, what size, what type? Are treads better across any type of terrain? They are the obvious choice for tough terrains, but they can complicate the mechanical design. Large wheels with four-wheel drive are also an obvious choice, but with larger wheels comes a decrease in the maximum available torque. Bipedal legs and legs that simulate other animals such as insects are much more complicated mechanical devices for mobile robots. They are good in narrow circumstances though. They are also good for tricky, inconsistent terrains, but they require a lot of servos and more importantly complex programming. The torque and speed capabilities depend on the motors they are using. Table 7.5 lists different types of robot mobility and the motor concerns for each.

Table 7.5 Types of Robot Mobility and Concerns for Each

Mobility Type	Features	Motor Concerns
Treads/tracks	Good for various terrains.	Require continuous rotation.
	Outdoor areas with tough terrains.	
	Constant contact with the ground prevents slipping.	
	Complicates mechanical design.	

Mobility Type	Features	Motor Concerns
Wheels	Good choice for beginners.	Require continuous rotation.
	Large wheels with four-wheel drive good for various terrains.	May require indirect drive mechanisms.
	Slippage possible.	Decrease maximum available torque.
Bipedal	Good for stairs and changing terrains.	Higher power demands.
	Complicated mechanical devices.	Require intricate positioning.
Bio-inspired legs	Good for tricky, inconsistent terrains.	Require many servos.
	Complex programming.	Require intricate positioning.

BRON'S Disaster and Recovery Challenges

Believe It Or Not!

At the DARPA Disaster and Recovery Challenge, each robot had to perform eight tasks. One of the tasks had to do with terrain and the other was a stairs task. Following are the logos for these tasks (see Figure 7.17).

LOGOS FOR RUBBLE & STAIRS DRC TASKS

RUBBLE   STAIRS

Figure 7.17
The logos used Rubble and Stairs tasks.

For the Rubble task, the robot earned one point if it successfully traversed either a debris field or the terrain field. The robot could have traversed both fields, but it would have still only earned one point.

The task was considered complete when all contacts between the terrain and the robot occurred in the when the robot was not in contact with any part of the terrain or debris.

The terrain contained cement blocks that were laid out, as much as possible and practical, so that any holes faced the side of the course rather than the start or end of the course. The blocks were not fastened to the ground, so they could shift during the run. For the debris side of the task, the debris laid directly between the start point and the finish. A robot had to get to the other side by either moving the debris or getting over it. The debris pieces were composed of lightweight components of less than 5 lbs.

For the Stairs task, the robot had to go up a flight of stairs. The stairway had a rail on the left side and no rail on the right side. The Stairs task was considered complete when all contact points lie on or above the top step.

Most of the teams opted for their robots to not perform the terrain task and perform the debris task by moving the debris as it moved from the starting to the end point. Robots that had wheels of some type performed better at this task than the bipedal robots. As humans, we can step decide to step over objects or even step on them and maintain our balance. This is a difficult task for a robot to perform. Some robots that were bipedal also had some type of wheeling mobility giving them the diversity to switch between the two. DRC-HUBO, the winner of the challenge, had wheels near its knee joints. It was able to transform from a standing position to a kneeling pose meant for wheeled and fast motion. Wheeled robots had difficulty performing the stairs challenge, as you can imagine. The most interesting performing this task was **CHIMP** (CMU Highly Intelligent Mobile Platform) and again DRC-HUBO. CHIMP had four legs like a chimp and wheels. It used its wheels in a most unusual way to navigate the stairs. HUBO was able to reverse the top part of its body, so it could navigate the stairs backwards. Following are some images of the robots performing these tasks (see Figure 7.18).

Figure 7.18
Robots performing Rubble and Stairs at the 2016 DARPA DRC Challenge.

ROBOTS PERFORMING FOR RUBBLE & STAIRS DRC TASKS

RUBBLE

STAIRS

MOMARO

DRC- HUBO

CHIMP

CHIMP

RUNNING MAN

RUNNING MAN

Torque Challenge for Robot Arm and End-Effectors

What about the motors for the robot and end-effectors? What are the torque and maximum rotation requirements needed? The actuators needed for a robot arm depend on the type of robot arm and the DOF (Degrees of Freedom). DOFs are associated with the arm joints that have a motor at each joint. Each joint has a maximum rotation and a startup torque. Determining whether the arm is up to the task requires calculations and a clear picture of what the arm is expected to do. The end-effectors are the hands and tools of the robot. They also have to be evaluated. This is discussed in the section "Robotic Arms and End-Effectors" later in this chapter.

Calculating Torque and Speed Requirements

How are the calculation formulas for speed, torque, and whatever used to help pick motors so the robot can perform the task? Torque and speed calculations are used to determine whether the robot can safely perform the task based on the capabilities of its motors. The amount of torque required by a robot to perform a task is called the *startup torque* and should be around 20% to 30% of its maximum torque (refer to Figure 7.6). It should not exceed the rated torque for the motor. For example, say a robot must accelerate to a speed of 1 cm/s. The attributes for the potential robots to perform the task are listed in Table 7.6.

Table 7.6   Torque and Speed Evaluation Between a Small and Larger Robot

Attributes	Small Robot	Larger Robot
Weight (m)	1 kg	4.5 kg
Wheel radius (r)	1 cm	4 cm
Motor type	Servo	DC Motor
No load speed (RPM)	170 RPM	146 RPM
No load torque (Stall)	5 kg/cm	20 kg/cm
Rated torque	2 kg/cm	12 kg/cm
Desired acceleration (a)	1.0 m/s^2	1.0 m/s^2
Desired speed (ds)	30 cm/s	30 cm/s
C = m * a * r (Startup torque)	1 * 1 * 1 = 1.0 Nm = 0.10 kg/cm	4.5 * 1 * 4 = 18 Nm = 1.8 kg/cm
RPM calculation: (1) Distance_per_rev = 2 * 3.141 * r (2) RPM = (ds/distance_per_rev) * 60	2 * 3.141 * 1 = 6.282 cm Desired speed = 30 cm per sec. 30/6.282 = 4.8 * 60 = 287 RPM	2 * 3.141 * 4 = 25.1 cm Desired speed = 30 cm per sec 30/25.1 = 1.2 * 60 = 72 RPM
Torque evaluation 20% to 30 % of rated torque	0.10 kg/cm / 2 kg/cm = 5% PASSED	1.8 kg/cm / 12 kg/cm = 15% PASSED
Speed evaluation	287 RPM > 170 RPM NOT PASSED	72 RPM < 146 RPM PASSED

There are two robots to choose from to execute this task. There is a formula that relates the motor torque with the robot's weight and acceleration desired. So this shows the torque needed for the task, the startup torque. This value should be no more than 20% to 30 % of the rated torque of the possible motors. A motion formula can be used that relates the weight of the robot, torque, and acceleration:

C/r = m * a + Fattr

where m = mass of the robot (weight in kg), r = radius of the wheel, a = the acceleration, and C is the torque. Fattr is the attrition force between the wheel and the floor. This has to do with the terrain challenges discussed earlier and can be difficult to determine. So for simplicity, we represent this value as m * a:

C/r = 2 (m * a)

Assuming the robot has two motors (left and right wheels), we calculate half this torque:

C = m * a * r

If C, the startup torque, is more than 20% to 30% of the rated torque of the motor, the robot will not be able to meet the desired acceleration. In this case, both the robot's motors meet the torque requirement. The torque evaluation in Table 7.6 shows that the startup torque is only 5% of the small robot's rated motor and 15% of the larger robot's motor.

What about speed (RPM)? How fast can the robot to go? How can we calculate how many rotations per minute and how far per minute? To calculate the distance a wheel travels in one revolution, use this calculation:

*distance = 2 * 3.141 * r*

and the RPMs:

RPM = (ds / distance) * 60

In this case, the speed evaluation is a little different. The smaller robot's motor cannot reach the desired speed and the larger robot's motor can.

Motors and REQUIRE

All these characteristics and calculations should be used to help determine whether the robot is capable of executing a task. The REQUIRE (Robot Effectiveness Quotient Used in Real Environments) is a checklist that can be used to figure this out. Here, let's make it clear that the actuator is one factor in making this determination. The goal for each factor (sensors, actuators, end-effectors, and microcontrollers) is to be as close to 100% as possible. For the tasks for the motors, use the calculations to make sure each of the tasks can be executed by the motors used. If so, 100% can be used for the actuators. Table 7.8 compares the DC and servo motors based on their advantages and disadvantages.

Table 8.7 Advantages and Disadvantages of the DC and Servo Motors

Type of Motor	Features	Advantages	Disadvantages
DC	Some are gear heads Uses PWM duty cycle to reduce/control speed Used for any weight robot	A lot of variety Powerful Easy to interface Good for large robots	Too fast Requires a high current Expensive Hard to mount on wheels Complex controls (PWM) May require external encoder
Servo	Has DC motor included Uses a closed loop control to maintain speed/position Robots up to 5 pounds	A lot of variety Good for indoor robot speed, small robots Good positioning robot arm joints Cheap Easy to mount on wheels Easy interface Medium power requirement	Low weight capabilities Little speed control

Programming the Robot to Move

Programming the motor is controlling the motor. We discussed earlier how motors are controlled by sending signals/commands for speed and positioning the shaft. Motor shafts coupled to wheels, legs, or tracks cause the robot to move. For wheeled and track wheeled robots, there are some basic movements:

- Forward in a straight line

- Rotate

- Arcs

- Stop

These movements can be sequenced together so a mobile robot can perform some sophisticated maneuvering and traveling in an environment. Each type of movement requires some additional information for that movement to be executed. For example, to move a robot forward or backward in a straight line, the motor needs to know how far or for how long. A rotation requires angle and a direction, and so on. Here is a list:

- **Forward:** Time, heading, distance

- **Rotate:** Angle in degrees, direction

- **Arc:** Radius, direction

Some moves can also be coupled together, like arc-forward. These parameters can be positive or negative to cause clockwise or counterclockwise, left or right, forward or backward movements. And of course speed is a parameter for these moves. But how is a motor programmed to execute these moves? It depends on how many motors are involved and some other things.

One Motor, Two, Three, More?

The wheel drive of the robot is an important part of how the motors are programmed and affects the mobility performance. Just like a car, a wheeled robot can have two-wheel or four-wheel drive. With four-wheel drive, a motor drives each wheel. With two-wheel drive, the two motors are in the front or in the back. For a tracked wheeled robot, there may be just two gears with a motor on each and tracks wrapped around additional wheels. This is the wheel drive configuration for our robot depicted in Figure 7.19 along with some other wheel drive configurations.

Figure 7.19
Some wheel drive configurations

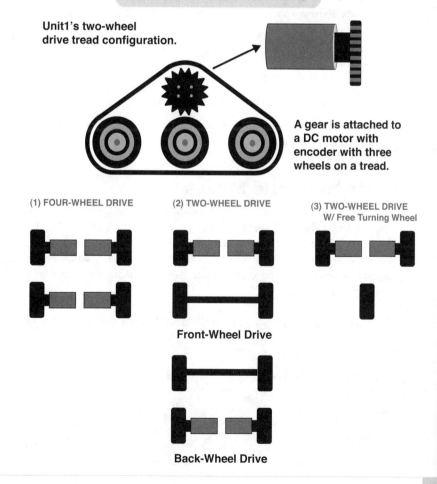

WHEEL DRIVE CONFIGURATIONS

Unit1's two-wheel drive tread configuration.

A gear is attached to a DC motor with encoder with three wheels on a tread.

(1) FOUR-WHEEL DRIVE

(2) TWO-WHEEL DRIVE

Front-Wheel Drive

Back-Wheel Drive

(3) TWO-WHEEL DRIVE
W/ Free Turning Wheel

Each type of wheel drive has advantages and disadvantages. More motors means more torque. We discussed earlier the various challenges for mobile robots, including the need for speed and torque to overcome weight, terrain, surface friction, and so on.

Making the Moves

There are different approaches to coordinate the motors to perform the forward, rotate, arc, and stop moves—for example, car-type steering called Ackerman steering, cab-drive steering, omni-directional wheels, and the differential steering method. The differential steering method is where the speed of the wheels is altered to change the direction the robot is moving or to turn the robot to some direction. Ackerman and the cab-drive steering has complex designs and servos and logic to control the robot. The simplest one for an introduction to programming autonomous robots is the basic differential steering method.

Differential steering creates different speeds and directions to perform the basic moves to make a robot mobile. With this method, there are typically two powered wheels, one on each side of the robot. Sometimes there are passive wheels that keep the robot from tipping over, like a castor wheel in the back. Here is a list of the basic operations for differential steering:

- When both wheels rotate at the same speed in the same direction, the robot moves straight in that direction (forward or backward).

- When one wheel rotates faster than the other, the robot turns in an arc toward the slower wheel.

- When the wheels rotate in opposite directions, the robot turns in place.

- When the two wheels on each adjacent side (left and right) rotate on opposite spin at the same speed, the robot spins 360°.

Programming the Moves

In this section, we will show BURT Translations for programming Tetrix DC motors using the EV3 microcontroller and the Tetrix Motor Controller. Listing 7.1 is the BURT Translation for programming the Tetrix DC motor using the leJOS API `TetrixControllerFactory`, `TetrixMotorController`, and `TetrixRegulatedMotor` classes. Listing 7.1 BURT Translation shows the pseudocode for testing motors and the Java code translation. This is the main line to test the motors by performing some basic operations.

BURT Translation Listing 7.1   Softbot Frame for `Unit1`'s Motor Test

BURT Translation INPUT

```
Softbot  Frame
Name:  Unit1
Parts:
Motor  Section:
Two DC motors
```

Actions:
Step 1: Initialize motors
Step 2: Test the motors by performing some basic operators

Tasks:
Test the DC motors by performing some basic operations.
End Frame

BURT Translations Output: Java Implementations

```
73    public static void main(String [] args)  throws Exception
74    {
75        basic_robot Unit1 = new basic_robot();
76        Unit1.testMotors();
77        Unit1.closeLog();
78    }
```

In line 75, the basic_robot is declared. Two functions/methods are then called. Listing 7.2 shows the code for the constructor.

Listing 7.2 Java Code for basic_robot **Constructor**

BURT Translations Output: Java Implementations

```
28    public basic_robot() throws InterruptedException,Exception
29    {
30        Log = new PrintWriter("basic_robot.log");
31        Port APort = LocalEV3.get().getPort("S1");
32        CF = new TetrixControllerFactory(APort);
33        Log.println("Tetrix Controller Factor Constructed");
34        MC = CF.newMotorController();
35        LeftMotor = MC.getRegulatedMotor(TetrixMotorController.MOTOR_1);
36        RightMotor = MC.getRegulatedMotor(TetrixMotorController.MOTOR_2);
37        LeftMotor.setReverse(true);
38        RightMotor.setReverse(false);
39        LeftMotor.resetTachoCount();
40        RightMotor.resetTachoCount();
41        Log.println("motors Constructed");
42        Thread.sleep(1000);
43    }
```

In line 31:

```
31    Port APort = LocalEV3.get().getPort("S1");
```

LocalEV3 is the instance of the EV3 microcontroller. This class is used to return various system resources connected to the microcontroller. This case, getPort(). returns the Port object for S1, the port where the DC motor controller is plugged in to. APort is passed to TetrixControllerFactory. In line 34:

```
34    MC = CF.newMotorController();
```

MC is a TetrixMotorController object, which is used to return instances of motor objects connected to it. In lines 35 and 36:

```
35    LeftMotor = MC.getRegulatedMotor(TetrixMotorController.MOTOR_1);
36    RightMotor = MC.getRegulatedMotor(TetrixMotorController.MOTOR_2);
```

LeftMotor and RightMotor are both TetrixRegulatedMotor objects. A regulated motor must have an encoder installed and connected to the controller for the methods to work properly. In lines 37 and 38:

```
37    LeftMotor.setReverse(true);
38    RightMotor.setReverse(false);
```

LeftMotor is set to go in reverse, and RightMotor is set to go forward. In lines 39 and 40:

```
39    LeftMotor.resetTachoCount();
40    RightMotor.resetTachoCount();
```

both motors' tachometers are reset. A tachometer is a device that measures the rotation speed of the motor's shaft measured in degrees. Resetting the tachometer causes the motor to stop rotating.

Listing 7.3 shows the BURT Translation for the testMotors() method and shows the pseudocode for testing motors and the Java code translation. This is the testMotors() method to test the motors by performing some basic operations.

BURT Translation Listing 7.3   Softbot Frame for Unit1's testMotors() Method

BURT Translation INPUT

```
Softbot  Frame
Name:  Unit1
Parts:
Motor  Section:
Two DC motors


Actions:
Step 1: Set speed of motors
Step 2: Rotate to the right
Step 3: Go forward
Step 4: Go backward
Step 5: Rotate to the left


Tasks:
Test the DC motors by performing some basic operations.
End Frame
```

BURT Translations Output: Java Implementations

```
48    public void testMotors() throws Exception
49    {
50        LeftMotor.setSpeed(300);
51        RightMotor.setSpeed(300);
52        LeftMotor.rotate(500,true);
53        RightMotor.rotate(-500);
54        Thread.sleep(2000);
55        LeftMotor.forward();
56        RightMotor.forward();
57        Thread.sleep(3000);
58        LeftMotor.backward();
59        RightMotor.backward();
60        Thread.sleep(2000);
61        RightMotor.rotate(500,true);
62        LeftMotor.rotate(-500);
63        Thread.sleep(2000);
64        LeftMotor.stop();
65        RightMotor.stop();
66    }
```

To program the motors to perform basic moves, each motor has to be controlled. To go forward based on the differential steering method, both wheels rotate at the same speed in the same direction; the robot moves straight in that direction (forward or backward).

In lines 50 and 51:

```
50    LeftMotor.setSpeed(300);
51    RightMotor.setSpeed(300);
```

both motors are set to the same speed. They are set to 300 deg/s2 (degrees per second squared). Setting or changing the speed on the motor requires an adjustment in the amount of power to the motor. The power value is derived from the passed value:

Power = Math.round((Value – 0.5553f) * 0.102247398f);

The actual speed value is not exact. The maximum reliably sustainable velocity for the Tetrix DC Gear motor is 154 RPM => 924 deg/s2.

To rotate the robot based on the differential steering method, wheels rotate in opposite directions; the robot turns in place.

In lines 52 and 53:

```
52    LeftMotor.rotate(500,true);
53    RightMotor.rotate(-500);
54    Thread.sleep(2000);
```

the motor rotates to the given Angle, measured in degrees to rotate the motor relative to the current position. LeftMotor is rotated to 500°, and RightMotor is rotated to –500°. This causes the robot to turn in place to the right. The true parameter is important. This means the method does not block but returns immediately so the next line of code can be executed. This is needed when coordinating the commands for both motors; they may have to execute their operations at the same time (close to it) to get the desired results. If true was not set, the LeftMotor would rotate before the RightMotor, causing a sloppy rotation; then the RightMotor would cause another sloppy rotation in the opposite direction. Thread.sleep() gives the robot time to perform the operation before executing the next commands.

Lines 55 and 56:

```
55    LeftMotor.forward();
56    RightMotor.forward();
```

cause both motors to go forward at the speed in the direction set in the constructor in lines 37 and 38:

```
37    LeftMotor.setReverse(true);
38    RightMotor.setReverse(false);
```

The motors can move forward, but they are set to move in opposite directions. How can they move forward? This is called *inverting motors*. How long will the robot move forward? Line 57:

```
57    Thread.sleep(3000);
```

is used for the time duration the robot is to move forward. In lines 58 and 59:

```
58    LeftMotor.backward();
59    RightMotor.backward();
```

the motors cause the robot to move backwards. Line 60:

```
60    Thread.sleep(2000);
```

is the time duration for this move.

In lines 61 and 62:

```
61    RightMotor.rotate(500,true);
62    LeftMotor.rotate(-500);
```

the motor rotates to the given Angle. LeftMotor is rotated to –500°, and the RightMotor is rotated to 500°. This causes the robot to turn in place to the left. In lines 64 and 65:

```
64    LeftMotor.stop();
65    RightMotor.stop();
```

both motors are stopped. Table 7.9 lists the classes used in Listings 7.1, 7.2, and 7.3.

Table 7.9 Classes Used in Listings 7.1, 7.2, and 7.3

Classes	Description	Methods Used
TetrixControllerFactory	Used to obtain motor and servo controller instances (TetrixMotorController and TetrixServoController).	newMotorController()
TetrixMotorController	Used to obtain motor and encoder instances TetrixMotor and TetrixEncoderMotor.	getRegulatedMotor()
TetrixRegulatedMotor	Tetrix DC motor abstraction with encoder support.	resetTachoCount() setReverse() setSpeed() rotate() forward() backward() stop()
LocalEV3	Instance of an EV3 device; used to obtain instances of resources connected to it	get() getPort()

Programming Motors to Travel to a Location

Mobile robots not only move around, forward, backward, and spin in place, they have things to do and places to be. They need to travel to a location to perform the task for a given situation. Now, this can be done by programming the motors as in Listing 7.3. It's a lot of work to coordinate the motors to move; there will be more work to coordinate the motors to take the robot to a specific location:

1. A path to the location has to be figured out—a path where the robot can move around, avoid obstacles, and so on.

2. While the robot is traveling, it has to know where it is and where it is heading at any given point. It may be necessary to correct its heading and to make sure it has not missed the target.

3. The path has to be translated into instructions or moves. Starting from the current location, the moves have to be determined to get the robot to the target.

4. The moves are then executed and the target is reached.

Figure 7.20 shows a flow of how these steps can be used to get a robot to a target location.

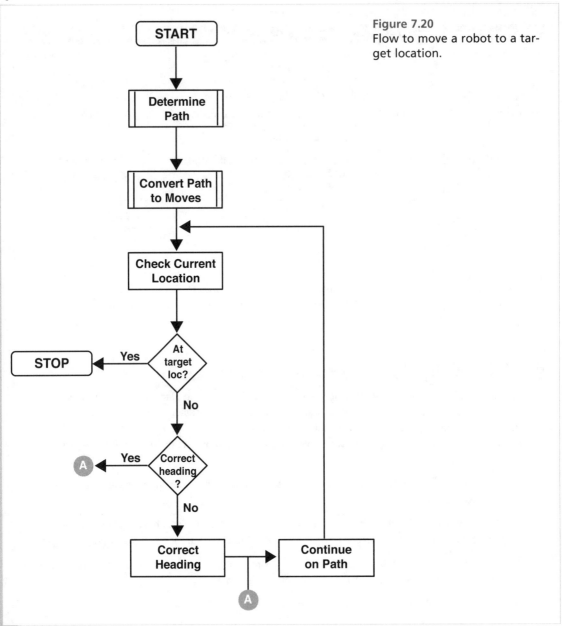

Figure 7.20
Flow to move a robot to a target location.

Listing 7.4 is the constructor for testing the motors' travel capabilities.

BURT Translation Listing 7.4 Unit1's Travel Test Constructor

BURT Translations Output: Java Implementations

```
37    public basic_robot() throws InterruptedException,Exception
38    {
39        Log = new PrintWriter("basic_robot.log");
40        WheelDiameter = 7.0f;
41        TrackWidth = 32.0f;
42
43        Port APort = LocalEV3.get().getPort("S1");
44        CF = new TetrixControllerFactory(APort);
45        Log.println("Tetrix Controller Factor Constructed");
46
47        MC = CF.newMotorController();
48        SC = CF.newServoController();
49        LeftMotor = MC.getRegulatedMotor(TetrixMotorController.MOTOR_1);
50        RightMotor = MC.getRegulatedMotor(TetrixMotorController.MOTOR_2);
51        LeftMotor.setReverse(true);
52        RightMotor.setReverse(false);
53        LeftMotor.resetTachoCount();
54        RightMotor.resetTachoCount();
55        Log.println("motors Constructed");
56        Thread.sleep(1000);
58        D1R1Pilot = new DifferentialPilot(WheelDiameter,
                                        TrackWidth,LeftMotor,RightMotor);
59        D1R1Pilot.reset();
60        D1R1Pilot.setTravelSpeed(10);
61        D1R1Pilot.setRotateSpeed(30);
62        D1R1Pilot.setMinRadius(30);
63        Log.println("Pilot Constructed");
64        Thread.sleep(1000);
65
66        CurrPos = new Pose();
67        CurrPos.setLocation(0,0);
68        Odometer = new OdometryPoseProvider(D1R1Pilot);
69        Odometer.setPose(CurrPos);
70        Log.println("Odometer Constructed");
71        Thread.sleep(1000);
72
73        D1R1Navigator = new Navigator(D1R1Pilot);
74        D1R1Navigator.singleStep(true);
75        Log.println("Navigator Constructed");
76        Thread.sleep(1000);
77    }
```

In the constructor, several new objects are declared. In lines 40 and 41:

```
40    WheelDiameter = 7.0f;
41    TrackWidth = 32.0f;
```

`WheelDiameter`, `TrackWidth`, `LeftMotor`, and `RightMotor` are parameters used to declare a `DifferentialPilot` object in line 58. `DifferentialPilot` defines methods that control robot movements such as travel forward or backward in a straight line or in a circular path, or rotate to a new direction.

The `DifferentialPilot` class only works with a wheel drive with two independently controlled motors, so a rotation turns the robot in place. `WheelDiameter` is the diameter of the wheel, and `TrackWidth` is the distance between the center of the right tire and the center of the left tire.

The following methods:

```
50    D1R1Pilot.reset();
60    D1R1Pilot.setTravelSpeed(10);
61    D1R1Pilot.setRotateSpeed(30);
62    D1R1Pilot.setMinRadius(30);
```

are all self-explanatory. All these methods sets the tachometer, travel speed, rotation speed, and minimum radius for both motors.

Line 68:

```
68    Odometer = new OdometryPoseProvider(D1R1Pilot);
```

declares an `OdometryPoseProvider` object. `OdometryPoseProvider` keeps track of the location and heading of the robot. A `Pose` object represents the location and heading (direction angle) of a robot. This class includes methods that update the `Pose` as the robot moves. All directions and angles are in degrees. The direction 0 is parallel to the X axis, and direction +90 is parallel to the Y axis. The `Pose` is passed to the `OdometryPoseProvider` in line 69:

```
69    Odometer.setPose(CurrPos);
```

In lines 73 and 74:

```
73    D1R1Navigator = new Navigator(D1R1Pilot);
74    D1R1Navigator.singleStep(true);
```

the `Navigator` object is declared. `Navigator` controls a robot's path traversal. If given multiple x,y locations (waypoints), the robot traverses the path by traveling to each waypoint in the order assigned. The `singleStep()` method controls whether the robot stops at each waypoint.

Listing 7.5 is the BURT Translation for the `testMoveTo()` method and shows the pseudocode for testing travel capabilities and the Java code translation. This is the `testMoveTo()` method to test the motors by performing some basic operations.

BURT Translation Listing 7.5   Softbot Frame for Unit1's testMoveTo() Method

BURT Translation INPUT

```
Softbot  Frame
Name:  Unit1
Parts:
Motor  Section:
Two DC motors


Actions:
Step 1: Report the current location of the robot
Step 2: Navigate to a location
Step 3: Travel a distance
Step 4: Rotate
Step 5: Report the current location of the robot


Tasks:
Test the traveling capabilities by performing some basic operations.
End Frame
```

BURT Translations Output: Java Implementations

```
80      public void rotate(int Degrees)
81      {
82          D1R1Pilot.rotate(Degrees);
83      }
84
85      public void forward()
86      {
87          D1R1Pilot.forward();
88      }
89
90      public void backward()
91      {
92          D1R1Pilot.backward();
93      }
94
95      public void travel(int Centimeters)
96      {
97          D1R1Pilot.travel(Centimeters);
98      }
99
100     public Pose odometer()
101     {
102         return(Odometer.getPose());
103     }
104
```

```
105     public void navigateTo(int X,int Y)
106     {
107         D1R1Navigator.clearPath();
108         D1R1Navigator.goTo(X,Y);
109     }
110
111     public void testMoveTo() throws Exception
112     {
113         Log.println("Position: " + odometer());
114         navigateTo(100,200);
115         Thread.sleep(6000);
116         travel(30);
117         Thread.sleep(1000);
118         rotate(90);
119         Thread.sleep(1000);
120         Log.println("Position: " + odometer());
121     }
```

Line 113:

```
113     Log.println("Position: " + odometer());
```

reports the current location of the robot. The odometer() method:

```
100     public Pose odometer()
101     {
102         return(Odometer.getPose());
103     }
```

calls the OdometerPoseProvider getPose() method, which returns a Pose object, which is reported to the log, for example:

Position: X:30.812708 Y:24.117397 H:62.781242

x, y, and heading are represented as floats.

In lines 114 and 116:

```
114     navigateTo(100,200);
116     travel(30);
```

are two methods to move a robot to a target location. The navigateTo(100,200) method:

```
105     public void navigateTo(int X,int Y)
106     {
107         D1R1Navigator.clearPath();
108         D1R1Navigator.goTo(X,Y);
109     }
```

is a wrapper for two `Navigator` methods. The `clearPath()` method clears any current paths. The `goTo()` method moves the robot toward the x, y location. If no path exists, a new one is created. The `travel()` method:

```
95    public void travel(int Centimeters)
96    {
97        D1R1Pilot.travel(Centimeters);
98    }
```

is a wrapper for the `DifferentialPilot` `travel()` method and moves the robot in a straight line the given centimeters. The robot travels forward if the value is positive and backward if the value is negative. When programming the motors to travel forward, each motor has to be given a forward command but no way to specify the distance.

Other methods are wrappers for the `DifferentialPilot` methods:

```
backward()
forward()
```

These methods move the robot forward or backward in a straight line. These methods perform similarly to the forward and backward methods for `TetrixRegulatedMotor`, but they move both motors. Table 7.10 lists the classes used in Listings 7.4 and 7.5.

Table 7.10 Classes Used in Listings 7.4 and 7.5

Classes	Description	Methods Used
DifferentialPilot	Used to control robot movements; methods include travel, forward, backward, rotate.	reset() setTravelSpeed() setRotateSpeed() setMinRadius() rotate() forward() backward() travel()
OdometerPoseProvider	Keeps track of the robot's location and heading.	setLocation()
Pose	Holds the x, y location and heading of a robot.	setPose() getPose()
Navigator	Controls the path a robot traverses, constructs a path based on waypoints.	singleStep() clearPath() goto()

Programming Motors Using Arduino

In this section, we show BURT Translations for programming brushless motors using the Arduino Uno microcontroller and Servo.h library. Listing 7.6 is the BURT Translation for programming the brushless (hobby servo) motor using 5V line. Listing 7.6 also shows the pseudocode for testing the motor and the C++ code translation.

BURT Translation Listing 7.6   Softbot Frame for Servo Motor Positioning Test

BURT Translation INPUT

```
Softbot  Frame
Name:  ServoMotor
Parts:
Motor  Section:
1 standard brushless servo motor

Actions:
Step 1: Initialize motor
Step 2: Position the servo from 0 to 180º in 5º increments
Step 3: Position the servo from 180 to 0º in 5º increments

Tasks:
Test the servo motor by controlling its position.
End Frame
```

BURT Translations Output: Java Implementations

```
1      #include <Servo.h>
2
3      Servo ServoMotor;
4
5      int Angle = 0;
6
7
8      void setup()
9      {
10         ServoMotor.attach(9);
11     }
12
13     void loop()
14     {
15         for(Angle = 0; Angle < 180; Angle += 5)
16         {
17             ServoMotor.write(Angle);
18             delay(25);
19         }
```

```
20
21          for(Angle = 180; Angle >= 5; Angle -= 5)
22          {
23              ServoMotor.write(Angle);
24              delay(25);
25          }
26      }
```

In line 10 in the setup() function, the attach() function connects the servo on pin 9 to the ServoMotor object. Certain versions of the Servo library support servos only on two pins: 9 and 10. In the loop() function, there are two for loops. The first for loop positions the angle of the servo from 0° to 180° in 5° increments. The second for loop positions the angle of the servo from 180° to 0° in 5° increments. The write() function sends a value to the servo that controls the shaft. If a standard servo is being used, this sets the angle of the shaft in degrees. If a continuous rotation servo is being used, this sets the speed of the servo with 0 being the full-speed of the servo in one direction, 180 being the full speed of the motor in the other direction, and 90 being no movement. The delays are used to allow time for the motor to position. Testing the speed of a continuous motor is done in Listing 7.7.

Listing 7.7 BURT Translation shows the pseudocode for testing two continuous servo motors' speed.

BURT Translation Listing 7.7   Softbot Frame for Two Servo Motors' Speed Test

BURT Translation INPUT

```
Softbot  Frame
Name:  ServoMotor
Parts:
Motor  Section:
2 continuous brushless servo motor

Actions:
Step 1: Initialize motors
Step 2: Rotate servos in the same direction from 90 to
        180º in 5º increments to increase speed
Step 3: Rotate servos in the same direction from 180 to
        90º in 5º increments to decrease

Tasks:
Test the servo motors by controlling their speed.
End Frame
```

BURT Translations Output: Java Implementations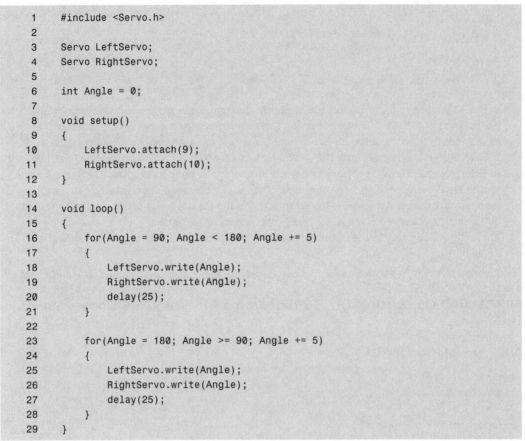

```
1    #include <Servo.h>
2
3    Servo LeftServo;
4    Servo RightServo;
5
6    int Angle = 0;
7
8    void setup()
9    {
10       LeftServo.attach(9);
11       RightServo.attach(10);
12   }
13
14   void loop()
15   {
16       for(Angle = 90; Angle < 180; Angle += 5)
17       {
18           LeftServo.write(Angle);
19           RightServo.write(Angle);
20           delay(25);
21       }
22
23       for(Angle = 180; Angle >= 90; Angle += 5)
24       {
25           LeftServo.write(Angle);
26           RightServo.write(Angle);
27           delay(25);
28       }
29   }
```

In Listing 7.7, there are two continuous rotating servos: LeftServo and RightServo. The write() function in this case causes the motors to increase or decrease their speed or stop rotating. In the first for loop, the control variant starts at 90°; the motors are not moving. Angle increases in 5° increments until it reaches 180°, the motors at full throttle. In the second for loop, the control variant starts at 180°; the motors are at full throttle. Angle decreases in 5° increments until it reaches 90°; the motors have stopped.

Robotic Arms and End-Effectors

Motors are also used to control a robotic arm and its end-effectors. Controlling a robotic arm means controlling each motor in the arm. Since the goal is to have intricate control for positioning, servo motors are used. Wherever there is a servo on the arm, this is considered a joint and a DOF. What defines a robot's motion capabilities is its degree of freedom, rotational range at each joint, and the dimension of space the arm works in, 2D or 3D. Movement can be determined by the rotation or a translation motion working in 2D or 3D space. The translational motion is how the joint can move forward and backward. This is defined by the type of robot arm being used. The end-effector of a robot arm is the device at the end of the robot. The end-effector is designed to interact with the environment and is some type of a gripper or a tool. That end-effector also is controlled by servos.

Robot Arms of Different Types

A robot arm has a length, joints, gripper, and maybe some type of tool. Figure 7.21 shows the basic components of a robot arm.

Figure 7.21
The basic components of a robot arm

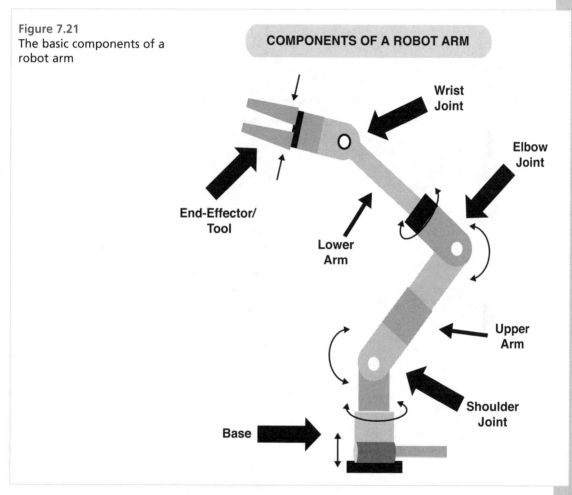

COMPONENTS OF A ROBOT ARM

Wrist
Joint

Elbow
Joint

End-Effector/
Tool

Lower
Arm

Upper
Arm

Shoulder
Joint

Base

A robot arm can be a standalone device with or without mobility, or it can be attached to a fully functioning robot. There are different types of robot arms, and each is characterized by

- Configuration space
- Workspace

The *configuration space* is the limitations of the robot arm's DOF. The workspace is the reachable space of the end-effector in 2D or 3D space. Figure 7.22 shows a few different types of robot arms and their configuration space and workspace.

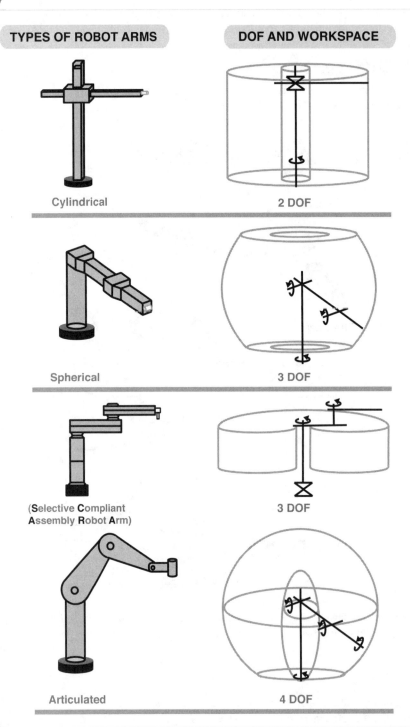

TYPES OF ROBOT ARMS

DOF AND WORKSPACE

Cylindrical

2 DOF

Spherical

3 DOF

(Selective Compliant Assembly Robot Arm)

3 DOF

Articulated

4 DOF

Figure 7.22
Different types of robot arms and their configuration space and workspace

Each rotary joint has restrictions or limitations to its movements based on the rotational range of the servo at that joint. But how much rotation is needed depends on the design or the location of the arm and the application. The translational motions are forward, backward, vertical, or horizontal. The sum of all the configuration spaces for all the joints of the arm, lengths of links (one joint to the next), the angle objects can be lifted, and so on define the *workspace*. It determines the fixed boundaries of the possible locations of the end-effector. Outside the workspace is where the end-effector cannot reach. A mobile robot can include the area the robot could physically relocate to. Inverse and forward kinematics is the method of figuring out the orientation and position of joints and end-effectors given the link lengths and joint angles. A discussion on kinematics is presented briefly at the end of this chapter.

Torque of the Robot Arm

Calculating the torque of the robot arm boils down to the combined torque of each servo in the arm. Just like calculating the torque of the DC motors to determine whether it was capable of moving the robot, the torque of the arm must be enough to lift a desired object, maybe carry or manipulate it. But now multiple servos, lengths of links, pivot points, and so on have to be considered. Figure 7.23 labels weights, joints, and lengths of links for a robot arm with 2 DOF used to lift an object.

Figure 7.23
Labeling of a 2 DOF robot arm.

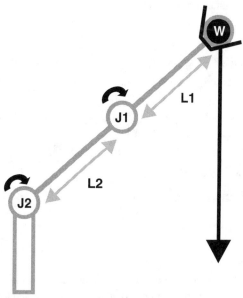

LABELING 2 DOF ROBOT ARM

Length of L1 + Length of L2 = Length of Arm
W = Weight of Object
J1 and J2 are joints.

The torque of each servo is calculated. To do this the robot arm has to be stretched to its maximum length; this is the maximum torque required. Also, the weight and length of each link, weight of each joint, and weight of the object to be lifted have to be figured out. The center of mass of each link is Length/2. Here are the calculations:

Torque About Joint A:

T1 = L1 / 2 * W1 + L1 * W4 + (L1 + L2 / 2) * W2 + (L1 + L3) * W3

Torque About Joint B:

T2 = L2 / 2 * W1 + L3 * W3

Each DOF added makes these calculation more complicated. The more DOF, the higher the torque requirements. Again, this torque calculation is the amount of torque needed for each joint. It should be well below its maximum torque. Table 7.11 shows the calculation performed on a 2 DOF robot arm. Based on this calculation, T2 has enough torque, but T1 does not.

Table 7.11 Robot Arm Torque Calculation Performed on a 2 DOF Robot Arm

Attributes	Robot Arm
DOF (w/o end-effector)	2
Length of arm L_{arm}	20 cm
Weight of beaker w/o substance W_{obj}	.297 kg
Dimensions of beaker	Circumference 24 cm
	Diameter 7.5 cm
	Height 9 cm
Weight of arm W_{arm}	.061 kg
Max torque (including additional gear sets)	Servo 1 5 kg cm
	Servo 2 6 kg cm
T1 = L1 / 2 * W1 + L1 * W4 + (L1 + L2 / 2) * W2 + (L1 + L3) * W3	T1 = (10 cm / 2 * .0305 kg) + (10 cm * .021 kg) + (10 cm + 10 cm / 2) * .0305 kg + (10 cm + 15 cm) * .297 kg
T2 = L2 / 2 * W1 + L3 * W3	T1 = (.1525 kg cm) + (.21 kg cm) + (.4575 kg cm) + (7.425 kg cm) = 8.245 kg cm
	T2 = 10 cm / 2 * .0305 kg + 15 cm * .297 kg
	T2 = .1525 kg cm + 4.455 kg cm = 4.6075 kg cm
Torque comparison	T1 = 8.245 kg cm > 5 kg cm NOT PASSED
	T2 = 4.6075 kg cm < 6 kg cm PASSED

For a robot arm that is purchased, this information should be supplied. If the torque required is determined, it can be compared to the stats of the arm. Table 7.12 contains the specification for the PhantomX Pincher robot arm used for Unit1.

Table 7.12 The Specification for PhantomX Pincher Robot Arm

Attributes	Stats
Weight	.55 kg
Vertical reach	35 cm
Horizontal reach	31 cm
Strength	25 cm /.04 kg
	20 cm/.07 kg
	15 cm/.10 kg
Gripper strength	.5 kg Holding
Wrist strength	.25 kg

Different Types of End-Effectors

End-effectors are designed to interact with the environment. They are what deliver the goods and do whatever task is to be performed. The arm delivers the end-effector to the point in space where it needs to be at the right time. There are many types of end-effectors:

- Mechanical grippers
- Negative pressure (vacuum)
- Magnetic
- Hooks
- Ladles (scoops liquid or powder)
- Others (electrostatic)

We used mechanical type end-effectors for the two robot arms. RS Media also has robot arms and end-effectors of a mechanical type. These are the most common. Mechanical grippers can be classified as:

- Parallel grippers
- Angular gripper
- Toggle gripper

Figure 7.24 shows mechanical type end-effectors used for our two robots.

Table 7.13 shows the robot arm and end-effector types for the Unit1 and Unit2 robots.

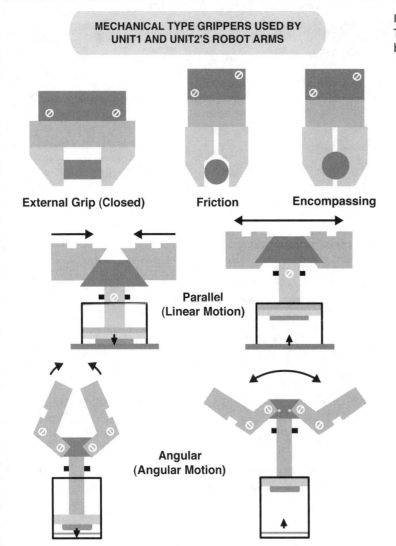

MECHANICAL TYPE GRIPPERS USED BY
UNIT1 AND UNIT2'S ROBOT ARMS

External Grip (Closed)   Friction   Encompassing

Parallel
(Linear Motion)

Angular
(Angular Motion)

Figure 7.24
Types of end-effectors used
by Unit1 and Unit2 robots

Table 7.13 The robot arms and end-effector types used for Unit1 and Unit2's robot arms.

ROBOT ARMS	ARM TYPES	END-EFFECTOR TYPES
Unit1 Robot Arm 1	Articulated (4 DOF)	Parallel (1 DOF)
		Mechanical gripper (External grip)
Unit1 Robot 2	Spherical (1 DOF)	Angular (1 DOF)
		Mechanical gripper (Friction)/Encompassing)
Unit2	Articulated (2 DOF)	Angular (3 DOF)
		Mechanical gripper (Encompassing)

Once it has been determined that the torque of the arm can lift the object, the end-effector should be evaluated. Just because the arm's servos can lift the object does not mean that the end-effector can grip and hold the object. To determine whether the end-effectors of a robot can be used to transport an object, a calculation can be performed to determine whether it has the force to hold the object:

$$F = \mu * W_{obj} * n$$

where F is the force required to hold the object, μ is the coefficient of the friction, W_{obj} is the weight of the object, and n is the number of fingers on the gripper.

But the torque of the gripper is not the sole factor. Other aspects of the gripper have to be considered. Look at the end-effectors of the three robot arms we used. Gripping implies there are mechanical fingers of some type. RS Media has hands, but they were not constructed to hold anything the size of the beaker. Robot arm 1 is the PhantomX Pincher Arm with 5 DOF including the end-effector. Based on its stats in Table 7.12, the gripper could hold the beaker:

.297 kg < .5 kg

but not at the wrist:

.297 kg > .25 kg

The other issue is the diameter of the beaker. The grip of the end-effector does not open wide enough. Unit1's robot arm 2 end-effector can hold the beaker as depicted in Figure 7.25.

UNIT1 AND UNIT2's ROBOT ARMS AND END EFFECTORS

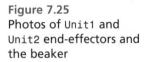

Figure 7.25
Photos of Unit1 and Unit2 end-effectors and the beaker

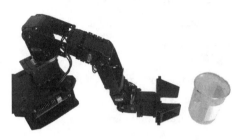

Unit1's Robot Arm 1 (PhantomX Pincher Robot Arm) end effector cannot hold/grasp the beaker.

Unit1's Robot Arm 2 end effector can hold/grasp the beaker.

Unit2's Robot Arm end effector was not required to hold or grasp anything in this scenario.

Programming the Robot Arm

In this section, we show the constructor for robot arm 2 (see Figure 7.26) using the EV3 microcontroller and Tetrix servo motors using the leJOS API `TetrixServoController` and `TetrixServo` classes (see Listing 7.8). Robot arm 2 has one DOF and a single servo to open and close the gripper.

Figure 7.26
Unit1's robot arm 2

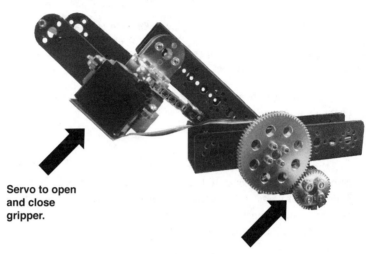

ROBOT ARM 1 DOF AND SERVO FOR GRIPPER

Servo to open
and close
gripper.

Servo connected to
pinion for 1 DOF.

BURT Translation Listing 7.8 Unit1's **Constructor for Servos for the Robot Arm and Gripper**

BURT Translations Output: Java Implementations

```
 9        TetrixServoController SC;
...
12        TetrixServo  Arm;
13        TetrixServo  Gripper;

31    public basic_robot() throws InterruptedException,Exception
32    {
33
...
49        Gripper = SC.getServo(TetrixServoController.SERVO_2);
50        Arm = SC.getServo(TetrixServoController.SERVO_1);
51        Log.println("Servos Constructed");
52        Thread.sleep(3000);
53        SC.setStepTime(7);
54        Arm.setRange(750,2250,180);
55        Arm.setAngle(100);
56        Thread.sleep(3000);
57        Gripper.setRange(750,2250,180);
58        Gripper.setAngle(50);
```

```
59          Thread.sleep(1000);
...
84      }
```

Lines 9, 12, and 13, and lines 49 through 59 have been added to the constructor. In lines 49 and 50:

```
49      Gripper = SC.getServo(TetrixServoController.SERVO_2);
50      Arm = SC.getServo(TetrixServoController.SERVO_1);
```

Gripper and Arm are assigned the TetrixServo objects for the servos for the arm and gripper. Line 53:

```
53      SC.setStepTime(7);
```

sets the "step time" for all servos plugged in to this controller. The step time is a delay before going on to the next command to be executed. This is similar to the sleep() methods, but where sleep() just waits until a command finishes executing, a step actually slows down the execution of a command or function. The value of steps is between 0 to 15. The isMoving() method always returns false if the step time is set to zero.

These methods:

```
54      Arm.setRange(750,2250,180);
55      Arm.setAngle(100);
        ...
57      Gripper.setRange(750,2250,180);
58      Gripper.setAngle(50);
```

are the workhorse methods for the arm and gripper. As discussed earlier in this chapter, the duration of the pulse determines the rotation of the servo motor. The setPulseWidth() method can be used to set the position of the robot arm. The width of the pulse must be within the maximum angle range of the servo. This method uses an absolute pulse and not a "relative" pulse width measured in microseconds. The parameter is between 750 to 2250 microseconds with a step resolution of 5.88 microseconds. A neutral position is 1500 microseconds width, the midpoint range of the servo.

The setRange() method sets the pulse width within the range of the servo in microseconds and the total travel range. For example, a 180° or 90° servo is the total travel range for the Hitec servos. The default total range is 200° with the low at 750 and high at 2250. This information must reflect the actual empirical specifications of a servo to be able to position the servo accurately. It accepts three parameters:

- microsecLOW: The low end of the servo response/operating range in microseconds

- microsecHIGH: The high end of the servo response/operating range in microseconds

- travelRange: The total mechanical travel range of the servo in degrees

In these cases, the arm and gripper servos have a range of 180°, taking advantage of the low and high microseconds for intricate positioning of the servos. The setAngle() method sets the angle target of the servo. Its accuracy depends on the parameters set in setRange().

These methods are used to set the position of the robot arm and the gripper. Listing 7.9 contains the methods to position both `Gripper` and `Arm` servo objects.

BURT Translation Listing 7.9   Methods to Move the `Gripper` and `Arm` **Servo Objects**

BURT Translations Output: Java Implementations

```
87      public void moveGripper(float X) throws Exception
88      {
89
90          Gripper.setAngle(X);
91          while(SC.isMoving())
92          {
93              Thread.sleep(1000);
94          }
95      }
96
97      public void moveArm(float X) throws Exception
98      {
99
100         Arm.setAngle(X);
101         while(SC.isMoving())
102         {
103             Thread.sleep(1000);
104         }
105     }
106
107
108     public void pickUpLargeObject() throws Exception
109     {
110         moveGripper(120);
111         moveArm(40);
112         moveGripper(10);
113         moveArm(100);
114     }
115
116     public void pickUpVeryLargeObject() throws Exception
117     {
118         moveArm(60);
119         moveGripper(120);
120         moveGripper(20);
121         moveArm(140);
122     }
123
124     public void putObjectDown() throws Exception
125     {
126         moveArm(10);
```

```
127         moveGripper(120);
128         moveArm(140);
129         moveGripper(10);
130     }
131
132     public void putLargeObjectDown() throws Exception
133     {
134         moveArm(40);
135         moveGripper(120);
136         moveArm(140);
137         moveGripper(10);
138     }
139
140     public void resetArm() throws Exception
141     {
142         moveArm(5);
143         moveGripper(10);
144     }
```

In Listing 7.9, the methods:

- moveGripper()

- moveArm()

both accept the angles to rotate the servos, which in turn positions the gripper and arm. The other methods:

- pickUpLargeObject()

- pickUpVeryLargeObject()

- putObjectDown()

- putLargeObjectDown()

- resetArm()

make calls to these methods, passing the desired angle.

Calculating Kinematics

Robot arm 2 has only 1 DOF and an end-effector with one servo. But the more DOFs, the more servos that have to be controlled. More DOFs means getting the end-effectors exactly where they are needed requires more work. How can you figure out where the servos should position the end-effector? If the servos are positioned at a certain angle, where is the end-effector positioned? Kinematics can be used to answer these questions.

Kinematics is a branch of mechanics used to describe the motion of points, objects, and groups of objects without considering what caused the motion. So kinematics can be used to describe the motion of the robot arm or the motion of the robot. Planar kinematics (PK) is motion in a plane or a

2D space. It can be used to describe the displacements of two points by means of rotation or translation. A robot moving across an environment is on a plane unless the robot falls off a table or a cliff. PK can be used to calculate the rotations of the motors used to move the robot on the plane. It can also be used to translate the position of an end-effector in 2D space (see Figure 7.27).

Figure 7.27
Planar kinematics for a robot and robot arm.

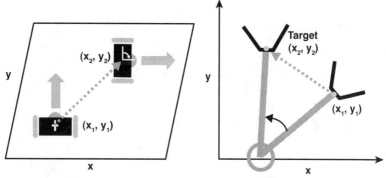

PLANAR KINEMATICS FOR TRANSLATION AND ROTATION

Mobile robot changes location and orientation.

1 DOF robot arm rotates to position end-effector at target location.

Kinematics used for robot arm manipulation is forward and inverse kinematics, but both areas have a 3D and 2D. We briefly discuss 2D (planar). Forward kinematics answers the first question:

How can we calculate the angles of joints to position the end-effector?

Forward kinematics is used to calculate the position of an end-effector from the angle positions of the joints. So for a robot arm with 2 DOF, where servo 1 (the shoulder) and servo 2, the elbow, are at some angle, where is the end-effector? Inverse kinematics is used to calculate the position of the joints (servo 1 and servo 2) based on the position of an end-effector. So with the same robot arm with 2 DOF, with the end-effector at a given position, what is the angle position of servo 1 (the shoulder), and what is the angle position of servo 2, the elbow? Figure 7.28 contrasts these two situations.

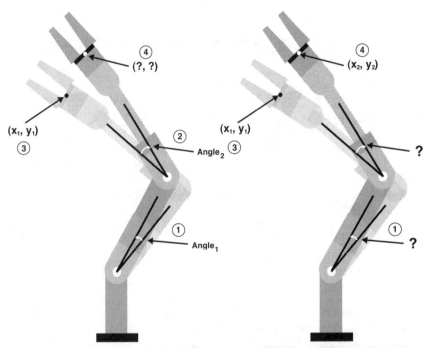

FORWARD VS. INVERSE KINEMATICS FOR 2 DOF ROBOT ARM

Figure 7.28
2 DOF robot arm contrasting forward and inverse kinematics

FORWARD KINEMATICS

1. **Angle$_1$** of shoulder from original to new location is **known**.
2. **Angle$_2$** of elbow from original to new location is **known**.
3. **Original x, y** location of end-effector is **known**.
4. **Determine new x, y** location of end-effector.

INVERSE KINEMATICS

1. **Determine Angle$_1$** of shoulder from original to new location.
2. **Determine Angle$_2$** of elbow from original to new location.
3. **Original x, y** location of end-effector is **known**.
4. **New x, y** location of end-effector is **known**.

Both are useful, but using them requires trigonometric and geometric equations. These equations can become complicated in a 3D space and where there are many joints. Once the equations are figured out, they must be translated into whatever language is being used to program the robot. Let's use planar kinematics to figure out the angles of the servo for the Robot arm 1. Figure 7.29 shows how the equations were derived.

So given the length of the arm and the desired location for the end-effector, the angle for the servo can be calculated. Listing 7.10 contains a BURT Translation to derive the angle for a servo. It shows the pseudocode and the partial C++ code translation.

Figure 7.29
1 DOF robot arm using inverse kinematics.

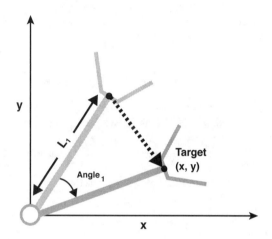

CALCULATING INVERSE KINEMATICS FOR 1 DOF ROBOT ARM

If $L_1 = \sqrt{x^2 + y^2}$ then location is reachable

$$\text{Angle}_1 = \text{atan2}(y, x)$$

BURT Translation Listing 7.10   Robot Arm 1 Inverse Kinematics

BURT Translation INPUT

```
Softbot  Frame
Name:  ServoMotor
Parts:
Motor  Section:
1 servo motor

Actions:
Given the length of the robot arm and the desired x,y location of
the end-effector,
Step 1: Square the x and y values
Step 2: If the Square root of the sum of squared x and y is less than
```

```
          the length of the arm
          Step 2:1: Take the arc tangent of x, y.
          Step 2:2: Use this value for the angle of the servo.

Tasks:
Test the servo motors by controlling their speed.
End Frame
```

BURT Translations Output: Java Implementations

```
      ...
300       SquaredX = Math.pow(X,2);
301       SquaredY = Math.pow(Y,2);
302       if(ArmLength >= Math.sqrt(SquaredX + SquaredY))
303       {
304           ArmAngle = math.atan2(X,Y);
305       }
306       ...
```

For a 2D 2 DOF robot arm, another procedure is used. Figure 7.30 shows how the equations are derived.

So given the length of the two links and the desired location for the end-effector, the angle for the servos can be calculated. The complete code listing for this BURT Translation can be viewed or downloaded from our website: www.robotteams.org/intro-robotics.

What's Ahead?

In this chapter, we talked about how to program different types of motors. In Chapter 8, "Getting Started with Autonomy: Building Your Robot's Softbot Counterpart," we will discuss how you start to program your robot to be autonomous.

Figure 7.30
2 DOF robot arm using
inverse kinematics

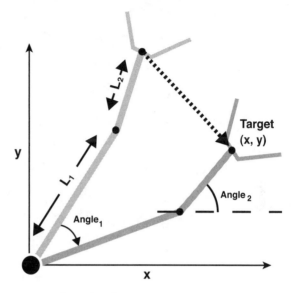

CALCULATING INVERSE KINEMATICS FOR 2 DOF ROBOT ARM

$$C = \frac{x^2 + y^2 - L^2 - L^2}{2L_1 L_2}$$

$$S^+ = \sqrt{1 - C^2}$$

$$\boxed{\text{Angle}_2^\pm = \text{atan2}\,(S^+, C)}$$

$$K_1 = L_1 + C\,L_2$$
$$K_2 = S^+ L_2$$

$$\boxed{\text{Angle}_2^\pm = \text{atan2}\,(y, x) - \text{atan2}\,(K_2, K_1)}$$

If $C^2 <= 1$ **then location is reachable.**

GETTING STARTED WITH AUTONOMY: BUILDING YOUR ROBOT'S SOFTBOT COUNTERPART

Robot Sensitivity Training Lesson #8: *It's not about the robot's hardware; it's "who" the robot is on the inside that matters.*

By definition, all real robots have some kind of end-effectors, sensors, actuators, and one or more controllers. While all these are necessary components, a collection of these components is not sufficient to be called a robot. These components can be used as standalone parts and can be put into many other kinds of machines and devices. It's how these components are combined, connected, and coordinated using programming that moves them in the direction of *robot*.

A robot is only as useful as its programming. Every useful autonomous robot has a *softbot* counterpart that ultimately gives the robot purpose, direction, and definition. For every programmable piece of hardware within or attached to a robot, there is a set of instructions that must be built to control it. Those instructions are used to program what functions each component performs given some particular input. Collectively, this set of instructions is the robot's softbot counterpart and captures the robot's potential behavior. For autonomous robots, the softbot counterpart controls the robot's behavior.

BRON'S
Believe It
Or Not!

Using Open Source Robot Hardware and Software Components Can Lead to Safer Robots

Open source robots are robots built using open source software components, robotics models, and hardware components. Open source robot software components give the programmer the option to inspect how a software component is designed or implemented prior to using it and to make any necessary safety improvements or adjustments. The more eyes that see the design and implementation of a piece of software, the more likely any hidden bugs or defects will be discovered and fixed. Transparency for robot operations is a good thing. Likewise, hardware designs and implementations that are open and accessible benefit from constructive criticism and knowledgeable community feedback.

The Bron Team recently tracked down Kyle Granat from Trossen Robotics. Trossen Robotics is one of the premier sources for accessible open source robots and robot components. They sell robot arms, hexapods, bipeds, along with Arduino-compatible controllers, sensors, and actuators. Some of their robot projects are shown in Figure 8.1.

TROSSEN ROBOT PROJECTS

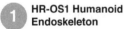

1 HR-OS1 Humanoid Endoskeleton

- 20 AX-12A Robot Actuators
- Arbotix-Pro Robocontroller
- Raspberry Pi 2 CPU with SD card
- Wifi Dongle and Bluetooth Dongle
- Open Source C++ framework
- Xbee, Wifi, Bluetooth

2 PhantomX Pincher Robot Arm

- AX-12A Dynamixel Actuators
- 5 DOF
- Arbotix Robocontroller for Onboard Processing
- Custom Parallel Gripper

3 PhantomX AX Metal Hexapod

- DYNAMIXEL Series Robot Servos
- 3 DOF Legs
- Arduino-Compatible ArbotiX
- Open Source Software
- Wireless Xbee Control
- Fully programmable for autonomy

Figure 8.1
Trossen Robotics HR-OS1 Humanoid Endoskeleton, PhantomX Pincher Robot Arm, and PhantomX AX Metal Hexapod

In addition to Kyle's Maker fair responsibilities, he wears many hats at Trossen as an engineer, shipper, sales, and even open source evangelist of sorts. According to Kyle, "Robots will be such a fundamental part of everyday life that robot programming in one form or another will be a basic skill of the next generation. ...Robot applications will be everywhere from trash collection to automotive construction. The availability and accessibility of a wide range of sensors and actuators all coordinated by an Arduino, opens the possibilities of robotics for anyone interested." And the prize doesn't always go to who has the most robot funding either. According to

Kyle, "When you don't have the funding to build the biggest and best; you find clever uses for parts that weren't necessarily designed to be used for a robot. ...In many cases you end up making the robot the best robot it can be when you have to be lean and frugal." Kyle suggests "that when building and learning how to program robots that there is an advantage to really understanding how the sensors, actuators, and controllers actually work rather than getting software and parts that hide how they operate." Simply because robotics applications are growing so fast and will eventually be everywhere and because robot applications can be built by anyone really interested, robot safety becomes an obvious issue. Having open standards and components like the ROS (Robot Operating System) will allow us to design more sophisticated systems with more and easier compatibility. According to Kyle, "The more open things are the more those protocols will be well defined. There is the ROS (robot operating system) which tries to have as many modules as possible so it tries to connect robot kits together."

 note

We use the term *softbot* because it helps us avoid the confusion over the robot's microcontroller versus the robot controller, or the robot's remote control.

The *softbot* plays the part of the robot controller. We use the term *softbot*, and that will help us avoid the confusion over the robot's microcontroller vs. the robot controller, or the robot's remote control. For our autonomous robots, the softbot is the set of instructions and data that control the robot. There are many approaches to building the softbot counterpart. The approaches range from explicitly programming every action the robot takes, to providing only reactive or reflexive programming, which allows the robot to choose its own actions as it reacts to its environment. So the robot can have a *proactive softbot*, a *reactive softbot*, or some hybrid combination of the two. This difference in softbot construction gives us three kinds of autonomous robots:

- Proactive autonomous robots

- Reactive autonomous robots

- Hybrid (proactive and reactive) autonomous robots

The range of completely proactive to completely reactive gives us the five basic levels of autonomy control shown in Table 8.1.

Table 8.1 Five Basic Levels of Robot Autonomy Control

Level	Programming	Autonomy Control
1	Explicit programming of all robot actions and functions.	Proactive
2	Explicit programing of some robot actions and usage of explicit planning algorithm for other actions and functions.	Proactive
3	Robot uses explicit planning algorithms for all actions and functions.	Proactive

Level	Programming	Autonomy Control
4	Robot uses some explicit planning algorithms and some reactive algorithms for actions and functions.	Hybrid (Proactive + Reactive)
5	Robot uses only reactive algorithms for actions and functions.	Reactive

 note

In this book, we show you how to build simple level 1 and level 2 softbots. For a detailed look at level 3 through level 5 control strategies, see *Embedded Robotics* by Thomas Braun and *Behavior-Based Robotics* by Ronald C. Arkin.

Softbots: A First Look

To see how this works, we use a simple robot build. Our simple robot, which we call Unit1, has the capability to move forward, backward, and turn in any direction. Unit1 only has two sensors:

■ **Ultrasonic sensor:** A sensor that measures distance

■ **16-color light sensor:** A sensor that identifies color

To get started, we created a scenario where there is a single object in the room with the robot. The robot's task is to locate the object, identify its color, and then report back. We want our robot to execute this task *autonomously* (i.e., without the aid of a remote control or human operator). The five essential ingredients for a successful robot autonomy are:

■ A robot with some capabilities

■ A scenario or situation

■ A role for the robot to play in the scenario or situation

■ Task(s) that can be accomplished with the robot's capabilities and that satisfy the robot's role

■ Some way to program the robot to autonomously execute the tasks

Table 8.2 The Five Essential Ingredients with Unit1

Ingredient Number	Ingredient (Required)	Ingredient Supplied
1	A robot with some capabilities.	Unit1 can move forward, backward, and turn. Unit1 has a: ■ Technic ultrasonic sensor. ■ Technic 16-color light sensor.
2	A scenario or situation.	A room with an object of unknown color.

Ingredient Number	Ingredient (Required)	Ingredient Supplied
3	A role for the robot to play in the scenario or situation.	Investigator.
4	Task(s) that can be accomplished with the robot's capabilities that satisfy the robot's role.	Locate the object, determine the color, and report.
5	Some way to program the robot to autonomously execute the tasks.	Unit1 has an EV3 microcontroller, running Linux, and can be programmed in Java.

Table 8.2 lists these five essential ingredients for a successful autonomous robot and who or what supplies each ingredient. Now that we are sure we have the five essential ingredients, the next step is to lay out the basic sections of Unit1's softbot frame. A softbot frame has at least four sections:

- Parts

- Actions

- Tasks

- Scenarios/situations

We describe these sections in simple English and then use our BURT translator to show what the simple English looks like translated into a computer language that supports object-oriented programming. In this simple example, we use the object-oriented language Java in the BURT translator. Listing 8.1 shows our first cut layout of Unit1's softbot frame.

Listing 8.1   First Cut Layout of Unit1 Softbot Frame

BURT Translation INPUT ⇒

```
Softbot  Frame
Name:  Unit1
Parts:
Sensor Section:
Ultrasonic Sensor
Light Sensor
Actuator Section:
Motors  with decoders (for movement)


Actions:
Step 1: Move forward some distance
Step 2: Move backward some distance
Step 3: Turn left some degrees
Step 4: Turn right some degrees
Step 5: Measure distance to object
```

```
Step 6: Determine color of object
Step 7: Report

Tasks:
Locate the object in the room, determine its color, and report.

Scenarios/Situations:
Unit1 is located in a small room containing a single object. Unit1 is playing the
role of an investigator and is assigned the task of locating the object, determining
its color, and reporting that color.

End Frame.
```

Parts Section

The Parts section should contain a component for each sensor, actuator, and end-effector the robot has. In this case, Unit1 only has two sensors.

The Actions Section

The Actions section contains the list of basic behaviors or actions that the robot can perform—for example, lifting, placing, moving, taking off, landing, walking, scanning, gliding, and so on. All these actions will be tied to some capability provided by the robot's build, its sensors, actuators, and end-effectors. Think about the basic actions the robot can perform based on its physical design and come up with easy-to-understand names and descriptions for these actions. An important part of robot programming is assigning practical and meaningful names and descriptions for robot actions and robot components. The names and descriptions that we decide on ultimately make up the *instruction vocabulary* for the robot. We discuss the instruction vocabulary and the ROLL model later in this chapter.

The Tasks Section

Whereas the Actions section is particular to the robot and its physical capabilities, the Tasks section describes activities specific to a particular scenario or situation. Like the Actions and Parts sections, the names and descriptions of tasks should be easy to understand, descriptive, and ultimately make up the robot's vocabulary. The robot uses functionality from the Actions section to accomplish scenario-specific activities listed in the Tasks section.

The Scenarios/Situations Section

The Scenarios/Situations section contains a simple (but reasonably complete) description of the scenario that the robot will be placed in, the robot's role in the scenario, and the tasks the robot is expected to execute. These four sections make up the basic parts of a softbot frame. As we see later, a softbot frame can occasionally have more than these four sections. But these represent the basic softbot skeleton. The softbot frame acts as a specification for both the robot and the robot's scenario. The softbot frame in its entirety must ultimately be translated into instructions that the robot's microcontroller can execute.

The Robot's ROLL Model and Softbot Frame

The softbot frame is typically specified using language from level 4 through level 7 (see Figure 8.2). During the specification of the softbot frame, we are not trying to use any specific programming language, but rather easy-to-understand names and descriptions relevant to the task at hand. The softbot frame is a design specification for the actual softbot and is described using the task vocabulary, the robot's basic action vocabulary, and terminology taken directly from the scenario or situation the robot will be placed in.

Figure 8.2
The robot's ROLL model introduced in Chapter 2

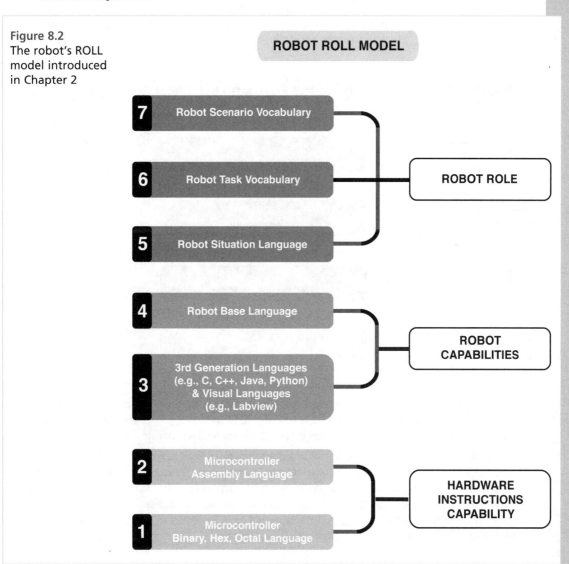

ROBOT ROLL MODEL

7	Robot Scenario Vocabulary
6	Robot Task Vocabulary
5	Robot Situation Language

ROBOT ROLE

| 4 | Robot Base Language |
| 3 | 3rd Generation Languages (e.g., C, C++, Java, Python) & Visual Languages (e.g., Labview) |

ROBOT CAPABILITIES

| 2 | Microcontroller Assembly Language |
| 1 | Microcontroller Binary, Hex, Octal Language |

HARDWARE INSTRUCTIONS CAPABILITY

Using language, vocabulary, and instructions from levels 4 through 7 allows you to initially focus on the robot's role, tasks, and scenario without thinking about the details of the microcontroller programming language or programming API. The softbot frame allows you to think about the robot's instructions from the point of view of the scenario or situations as opposed to the view of the microcontroller or any other piece of robot hardware.

Of course, the softbot frame must ultimately be translated into level 3 or level 2 instructions with some programming language. But specifying the robot and the scenario using language levels 4 through 7 helps to clarify both the robot's actions and the expectations of the role the robot is to perform in the scenario and situation.

We start with a level 4 through level 7 specification and we end with a level 3 (sometimes 2) specification, and the compiler or interpreter takes the specification to level 1. Figure 8.3 shows which ROLL model levels to use with the sections of the softbot frame.

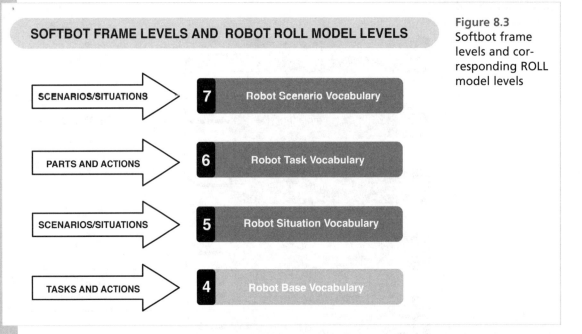

Figure 8.3
Softbot frame levels and corresponding ROLL model levels

We can develop our robot's language starting at the lowest level hardware perspective and moving up to the scenario or situation the robot will be used in, or we can approach the design of our robot language starting with the scenario down to the robot's lowest level hardware.

During the initial design phases of the robot's programming, it is advantageous to approach things from the scenario to the hardware. You should specify the softbot frame using names and descriptions that are meaningful and appropriate for levels 4 through 7 initially. Once the scenario is well understood and the expectations of the robot are clear, then we can translate levels 4 through 7 into a level 3 specification or a level 2 specification. In this book, we assume that you are responsible for each level of the robot's softbot frame specification.

In some robot applications the responsibilities are divided. The individual programming device drivers for the robot's parts may be different from the individual writing the level 3 robot instructions. The individual writing the level 3 instructions might not be responsible for writing the levels 4 through 7 specification, and so on. This is usually the case when a team of robot programmers is involved, or when robot software components originate from different sources. However, we describe the complete process starting with the high level (Natural-Language-Like) specification of the softbot frame using levels 4 through 7 to the translation of the softbot frame to Java or C++.

 tip

Notice for Ingredient #5 in Table 8.2 that Unit1 uses Java. In particular, Unit1 uses Java and the LeJOS Java class library running on Linux on the Mindstorms EV3 microcontroller. We use our BURT translator to show the translation between the softbot frame and Java.

BURT Translates Softbots Frames into Classes

The Tasks section of the softbot frame represents what is known as the agent loop, and the Parts, Actions, Situations, and Scenarios are implemented as objects and object methods. In Chapters 8 through 10, we present only introductory level material for objects and agents as it relates to robot programming. For a more detailed discussion of the subjects, see *The Art of Agent-Oriented Modeling* by Leon S. Sterling and Kuldar Taveter.

 note

It is important to note that each softbot frame is translated into one or more class specifications using an object-oriented language such as Java or C++. Technically, a softbot frame is an object and agent-oriented approach to specifying a robot controller.

A BURT Translation from Softbot Frame to Java Code

The BURT Translator consists of an Input section and an Output section. The Input section contains a natural language, or higher-level description of robot specifications and instructions, and the Output section contains a translation to a lower-level description of the robot specification or instructions. The Output level is always a lower-level than the Input level. The Input section might contain ROLL model level 4 instructions, and the Output section might contain the level 3 translation. The Input level could be level 5, and the Output level could be level 2, and so on. In some cases BURT shows a translation from one high level to another high level—for instance, level 6 instructions (robot task vocabulary) translated into level 4 (robot base vocabulary). The BURT Translation in Listing 8.2 shows an initial translation of Unit1's softbot frame to Java code.

Listing 8.2   BURT Translation of Unit1's Softbot Frame to Java

BURT Translation INPUT

```
Softbot  Frame
Name:  Unit1
Parts: SECTION 1
Sensor Section:
Ultrasonic Sensor
Light Sensor
Actuator Section:
Motors  with decoders (for movement)

Actions:  SECTION 2
Step 1: Move forward some distance
Step 2: Move backward some distance
Step 3: Turn left some degrees
Step 4: Turn right some degrees
Step 5: Measure distance to object
Step 6: Determine  color of object
Step 7: Report

Tasks: SECTION 3
Locate the object in the room, determine its color, and report.

Scenarios/Situations: SECTION 4
Unit1 is located in a small room  containing  a single object. Unit1 is playing the
role of an investigator and is assigned the task of locating the object, determining
its color, and reporting that color.

End Frame.
```

BURT Translations Output: Java Implementations

```java
class basic_ robot{

//PARTS: SECTION 1
// Sensor Section
   protected  EV3UltrasonicSensor  Vision;
   protected HiTechnicColorSensor ColorVision;
// Actuators
   protected TetrixRegulatedMotor LeftMotor;
   protected TetrixRegulatedMotor RightMotor;
   DifferentialPilot  D1R1Pilot;
//Situations/Scenarios: SECTION 4
   PrintWriter Log;
```

```java
    situation Situation1;
    location RobotLocation;


//ACTIONS: SECTION 2
  basic_robot()
  {
      Vision = new EV3UltrasonicSensor(SensorPort.S3);
      Vision.enable();
      Situation1 = new situation();
      RobotLocation = new location();
      RobotLocation.X = 0;
      RobotLocation.Y = 0;
      ColorVision = new HiTechnicColorSensor(SensorPort.S2);
      Log = new PrintWriter("basic_robot.log");
      Log.println("Sensors  constructed");
      //...
  }


  public void travel(int Centimeters)
  {
      D1R1Pilot.travel(Centimeters);
  }


  public int getColor()
  {
      return(ColorVision.getColorID());
  }


  public void rotate(int Degrees)
  {
      D1R1Pilot.rotate(Degrees);
  }


//TASKS: SECTION  3

  public void moveToObject() throws Exception
  {
      travel(Situation1.TestRoom.TestObject.Location.X);
      waitUntilStop(Situation1.TestRoom.ObjectLocation.X);
      rotate(90);
      waitForRotate(90);
      travel(Situation1.TestRoom.TestObject.Location.Y);
      waitUntilStop(Situation1.TestRoom.ObjectLocation.Y);

  }


  public void identifyColor()
  {
```

```java
        Situation1.TestRoom.TestObject.Color = getColor();
    }

    public void reportColor()
    {
        Log.println("color = " + situation1.TestRoom.TestObject.Color);
    }

    public void  performTask() throws Exception
    {
        moveToObject();
        identifyColor();
        reportColor();
    }

    public static void main(String [] args)  throws Exception
    {
        robot   Unit1 = new basic_robot();
        Unit1.performTask();
    }

}
```

The softbot frame is divided into four sections:

- Parts

- Actions

- Tasks

- Scenarios/situations

By filling out each of these four sections, you are developing a complete idea of:

- What robot you have

- What actions the robot can take

- What tasks you expect the robot to perform

- What scenario/situation the robot will perform the task in

Recall criterion #2 or our seven criteria for defining a true robot in Chapter 1, "What Is a Robot Anyway?":

Criterion #2 Programmable Actions and Behavior

There must be some way to give a robot a set of instructions detailing

- What actions to perform

- When to perform actions

- Where to perform actions

- Under what situations to perform actions

- How to perform actions

The four sections of the softbot frame allow us to completely describe the robot's programmable actions and behavior. Usually, if you do not have all the information to fill out these four sections, there is something about the robot's role and responsibilities in the given situation that is not understood, has not been planned for, or has not been considered. On the other hand, once these sections are complete, we have a roadmap for how to implement the robot's autonomy.

The specification of the softbot frame should not be in a programming language; it should be stated completely in a natural language like Spanish, Japanese, English, and so on. Some pseudocode can be used if needed to clarify some part of the description. Each of the four sections of the softbot frame should be complete so that there is no question what, when, where, and how the robot is to perform. Once the sections of the softbot frame are complete and well understood, we can code them in an appropriate object-oriented language. We specify object-oriented language for softbot frame implementation because classes, inheritance, and polymorphism all play critical parts in the actual implementation.

For Every Softbot Frame Section There Is an Object-Oriented Code Section

If we look at BURT Translation Listing 8.2, there is a Java code section for every softbot frame section. Although some of the low-level detail is left out of our initial draft, we have shown the Java code for all the major components of the softbot frame. BURT Translation Listing 8.2 shows the actual Java code uploaded to our robot for execution.

Section 1: The Robot Parts Specification   Section 1 contains the declarations of the components that the robot has: the sensors, motors, end-effectors, and communication components. Any programmable hardware component can be specified in Section 1, especially if that component is going to be used for any of the situations or scenarios the robot is being programmed for. For our first autonomous robot example, we have a `basic_robot` class with modest hardware:

- Ultrasonic sensor

- Light color sensor

- Two motors

Whatever tasks we have planned for the robot have to be accomplished with these parts. Notice that in the specification in the softbot frame, we only have to list the fact that the robot has an ultrasonic sensor and a light color sensor. And if we look at the BURT Translation of Section 1, we see the declarations of the actual parts and Java code for those parts:

```
protected EV3UltrasonicSensor Vision;
protected HiTechnicColorSensor ColorVision;
```

We have the same situation with the motors. In the softbot frame, we simply specify two motors with decoders, and that eventually is translated into the following appropriate Java code declarations:

```
protected TetrixRegulatedMotor LeftMotor;
protected TetrixRegulatedMotor RightMotor;
```

In this case, the sensor and motor components are part of the leJOS Library. The leJOS is Java-based firmware replaced for the LEGO Mindstorms robot kits. It provides a JVM-compatible class library that has Java classes for much of the Mindstorms robotic functionality. A softbot frame can be and is usually best off when it is platform independent. That way particular robot parts, sensors, and effectors can be selected after it is clear what the robot needs to actually perform. For instance, the implementation of the softbot frame could be using Arduino and C++, and we might have code like the following that is used by the Arduino environment:

```
Servo LeftMotor;
Servo RightMotor;
```

Once we have specified the softbot frame, it can be implemented using different robot libraries, sensor sets, and microcontrollers. In this book, we use Java for our Mindstorms EV3 NXT-based robot examples as well as for RS Modia robot examples, and the Arduino platform for our C++ examples.

Section 2: The Robot's Basic Actions   Section 2 has a simple description of the basic actions the robot can perform. We don't want to lists tasks that the robot is to perform in this section. We do want to list basic actions that are independent of any particular tasks and represent the fundamental capabilities of the robot, for example:

- Walk forward
- Scan left, scan right
- Lift arm, and so on

 tip

Look at the softbot frame as part of a design technique that allows you to design the set of instructions that your robot needs to execute in some particular scenario or situation, without having to initially worry about specific hardware components or robot libraries.

The BURT Translation Listing 8.1 shows that our `basic_robot` has seven basic actions:

- Move forward some distance
- Move backward some distance
- Turn left some degrees
- Turn right some degrees
- Measure distance to object
- Determine color of object
- Report (Log)

After specifying the robot's basic capabilities, we should have some hint to whether the robot will be able to carry out the tasks it will be assigned. For example, if our scenario required that the robot climb stairs, or achieve various altitudes, the list of actions that our `basic_robot` can perform would appear to come up short. The robot capabilities listed in the Actions section are a good early indicator of whether the robot is up to the task.

The tasks that the robot is assigned to perform must ultimately be implemented as a combination of one or more of the basic actions listed in the Actions section. The actions are then in turn implemented by microcontroller code. Figure 8.4 shows the basic relationship between tasks and actions.

Figure 8.4
Basic relationship between tasks and actions

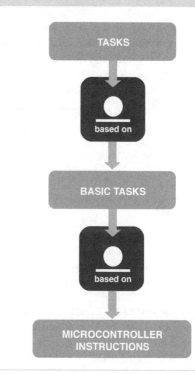

RELATIONSHIP BETWEEN TASKS AND ACTIONS

TASKS

based on

BASIC TASKS

based on

MICROCONTROLLER
INSTRUCTIONS

The Log action in Section 2 allows the robot to save its sensor values, motor settings, or other pieces of information that are accumulated during its actions and execution of tasks. There should always have to be a code counterpart for anything in the softbot frame, but there is not always a softbot frame component for a piece of code that is used. Take a look at the following code shown previously in Listing 8.2 Section 2 of the Java code containing the `basic_robot()` constructor:

```
basic_robot()
{
    Vision = new EV3UltrasonicSensor(SensorPort.S3);
    Vision.enable();
    Situation1 = new situation();
    RobotLocation = new location();
    RobotLocation.X = 0;
```

```
    RobotLocation.Y = 0;
    ColorVision = new HiTechnicColorSensor(SensorPort.S2);
    Log = new PrintWriter("basic_robot.log");
    Log.println("Sensors   constructed");
    //...
}
```

The softbot frame does not mention this action.

The constructor is responsible for the power-up sequence of the robot. It can be used to control what happens when the robot is initially turned on. This includes any startup procedures for any components, port settings, variable initializations, calibrations, speed settings, power checks, and so on. A lot of hardware and software startup housekeeping takes place when a robot is first powered on. This is best put in the constructor but can sometimes be put into initialization routines like the setup() procedure the Arduino programming environment uses. This level of detail can be put into the softbot frame depending on the level of autonomy design being specified. Here we omit it to keep our initial softbot frame as simple as practical.

Notice that the list of basic actions in Section 2 of the softbot frame are translated into implementations in the Action Code section. For example:

```
Softbot Frame Section 2:
Determine color of object
```

is ultimately translated into the following basic_robot Java method in Section 2 as a full implementation:

```
public int getColor()
{
return(ColorVision.getColorID());
}
```

Here the getColor() method uses the ColorVision object to scan the color of an object and return its color id. This code is the implementation of the getColor() instruction. Notice that the softbot frame descriptions and the name of Java code member functions and methods convey the same idea, for example:

```
Determine Color of Object and getColor()
```

This is not a coincidence. Another important use of the softbot frame is to give the robot programmer hints, indicators, and ideas for how to name routines, procedures, object methods, and variables. Remember the robot's ROLL model. We try to keep level 3 method, routine, and variable names as close as practical to the corresponding level 4 and level 5 descriptions. By comparing each section in the BURT Translation with its Java code equivalent, we can see how this can be accomplished.

Section 3: The Robot's Situation-Specific Tasks   Whereas the Actions section is used to describe the robot's basic actions and capabilities independent of any particular task or situation, the Tasks section is meant to describe tasks that the robot will execute for a particular scenario or situation.

 note

Actions are robot specific; tasks are situation/scenario specific.

In our sample scenario, the robot's tasks involve approaching some object and reporting its color. The softbot frame lists the tasks as:

```
Locate the object in the room, determine its color and report.
```

The BURT Translation shows the actual implementation for the three tasks:

```
Move to object
Determine its color
Report
```

Tasks are implemented using the basic_robot actions described in Section 2. Listing 8.3 shows the implementation of moveToObject() shown in the BURT Translation.

Listing 8.3   Definition of moveToObject() Method

BURT Translations Output: Java Implementations

```java
public void moveToObject() throws Exception
{
    travel(Situation1.TestRoom.TestObject.Location.X);
    waitUntilStop(Situation1.TestRoom.ObjectLocation.X);
    rotate(90);
    waitForRotate(90);
    travel(Situation1.TestRoom.TestObject.Location.Y);
    waitUntilStop(Situation1.TestRoom.ObjectLocation.Y);
}
```

Notice that the travel() and rotate() actions were defined in the Actions section and are used to help accomplish the tasks.

Synchronous and Asynchronous Robot Instructions   BURT Translation Listing 8.3 contains two other actions that we haven't discussed yet: waitUntilStop() and waitForRotation(). In some robot programming environments, a list of instructions can be considered totally synchronous (sometimes referred to as *blocking*). That is, the second instruction won't be executed until the first instruction has completed. But in many robot programming environments, especially robots consisting of many servos, effectors, sensors, and motors that can operate independently, the instructions may be executed asynchronously (sometimes referred to as nonblocking). This means that the robot may try to execute the next instruction before the previous instruction has finished. In Listing 8.3, the waitUntilStop() and waitForRotate() force the robot to wait until the travel() and rotate() commands have been completely executed. The moveToObject() does not take any special arguments, so the question might be asked, move to what object? Where? Look at the argument of the travel() function calls in Listing 8.3. They tell the robot exactly where to go:

```
Situation1.TestRoom.ObjectLocation.X
Situation1.TestRoom.ObjectLocation.Y
```

What is `Situation1`, `TestRoom`, and `TestObject`? Our approach to programming autonomous robots requires that the robot be instructed to play some specific role in some scenario or situation. The softbot frame must contain a specification for the scenario and situation.

Section 4: The Robot's Scenario and Situation   Section 4 of our softbot frame specifies the robot's scenario as:

Unit1 is located in a small room containing a single object. Unit1 is playing the role of an investigator and is assigned the task of locating the object, determining, and reporting its color.

The idea of specifying a particular situation for the robot to perform a particular task and play a particular role is at the heart of programming robots to act autonomously. Figure 8.5 shows three important requirements for autonomous robot program design.

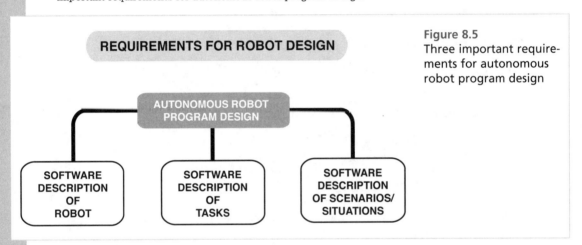

REQUIREMENTS FOR ROBOT DESIGN

AUTONOMOUS ROBOT PROGRAM DESIGN

SOFTWARE DESCRIPTION OF ROBOT

SOFTWARE DESCRIPTION OF TASKS

SOFTWARE DESCRIPTION OF SCENARIOS/ SITUATIONS

Figure 8.5
Three important requirements for autonomous robot program design

The BURT Translation Listing 8.2 shows the declaration necessary for our situation:

```
situation Situation1;
location RobotLocation;
```

However, it doesn't show the actual implementation. First, it is important to note the relationship between the `basic_robot` class and the robot's situation. For a robot to act autonomously, every robot class has one or more situation class as part of its design. We can implement model situations using object-oriented languages such as C++ and Java. So the `basic_robot` class is used to describe our robot in software, and a situation class is used to describe our robot's scenario or situation in software. We describe the process of capturing the detail of a scenario/situation in a class in detail in Chapter 9, "Robot SPACES." Let's take the first look at a simple definition of our robot's situation shown in Listing 8.4.

Listing 8.4 Definition of situation Class

BURT Translations Output: Java Implementations

```java
class situation{
   public room TestRoom;
   public situation()
   {
       TestRoom = new room();
   }
}
```

The situation class consists of a single data element of type room named Test room and a constructor that creates an instance of room. So BURT Translation Listing 8.4 shows us that our basic_robot class has a situation that consists of a single room. But what about the object that the robot is supposed to approach and determine the color for? It's not part of the situation. The object is located in the room. So, we also have a room class shown in BURT Translation Listing 8.5.

Listing 8.5 The room Class

BURT Translations Output: Java Implementations

```java
class room{
   protected int Length = 300;
   protected int Width = 200;
   protected int Area;
   public something TestObject;

   public  room()
   {
       TestObject = new something();
   }

   public int area()
   {
       Area = Length * Width;
       return(Area);
   }

   public int length()
   {
       return(Length);
   }

   public int width()
```

```
    {
        return(Width);
    }
}
```

The basic_robot has a situation. The situation consists of a single room. The room has a Length, Width, Area, and TestObject. The robot can find out information about the TestRoom by calling area(), length(), or width() methods. Also, the TestObject is part of the room class. The TestObject is also implemented as an object of type something. Listing BURT Translation Listing 8.6 shows the definitions of the classes something and location.

Listing 8.6 Definitions of the something and location Classes

BURT Translation Output: Java Implementations

```java
class location{
    public int X;
    public int Y;
}

class something{
    public int Color;
    public location Location;
    public something()
    {
        Location = new location();
        Location.X = 20;
        Location.Y = 50;
    }
}
```

These classes allow us to talk about the object and its location. Notice in Listing 8.6 that a something object will have a color and a location. In this case, we know where the object is located: coordinates(20,50), but we do not know what color the object is. It is Unit1's task to identify and report the color using its color sensor. So the situation, room, something, and location classes shown previously in BURT Translation Listing 8.4 through Listing 8.6 allow us to describe the robot's scenario/situation as Java code. Once we describe the robot, its tasks, and its situation in code, we can direct the robot to execute that code without the need for further human intervention or interaction.

The Java code in BURT Translation Listing 8.7 shows us how the robot will implement its tasks for the given situation.

BURT Translation Output: Java Implementation

Listing 8.7   BURT Translation for `moveToObject()`, `identifyColor()`,
`reportColor()`, **and** `performTasks()` **Methods**

```java
public void moveToObject() throws Exception
{
    travel(Situation1.TestRoom.TestObject.Location.X);
    waitUntilStop(Situation1.TestRoom.ObjectLocation.X);
    rotate(90);
    waitForRotate(90);
    travel(Situation1.TestRoom.TestObject.Location.Y);
    waitUntilStop(Situation1.TestRoom.ObjectLocation.Y);
}

public void identifyColor()
{
    Situation1.TestRoom.TestObject.Color = getColor();
}

public void reportColor()
{
    Log.println("color = " + situation1.TestRoom.TestObject.Color);
}

public void performTask() throws Exception
{
    moveToObject();
    identifyColor();
    reportColor();
}
```

To have the robot execute its task autonomously, all we need to do is issue the commands:

```java
public static void main(String [] args)   throws Exception
{
    basic_robot Unit1 = new basic_robot();
    Unit1.performTask();
}
```

and `Unit1` executes its task.

Our First Pass at Autonomous Robot Program Designs

Having `Unit1` approach an object and report its color was our first pass at an autonomous robot
design. In this first pass, we presented an oversimplified softbot frame, scenario, and robot task so

that you have some idea of what the basic steps and parts are to autonomous robot programming. But these oversimplifications leave many questions: What if the robot can't find the object? What if the robot goes to the wrong location? What happens if the robot's color sensor cannot detect the color of the object? What if there is an obstacle between the robot's initial position and the position of the object? How do we know the robot's initial position? In the next couple of chapters, we provide more details to our simplified softbot frame, and we expand on techniques of representing situations and scenarios using classes.

We ask Unit1 to not only report the object's color but retrieve the object as well. And we take a closer look at the scenario and situation modeling and how they are implemented in Java or C++ environments. Keep in mind that all our designs are implemented as Arduino, RS Media, and NXT-Mindstorms based robots. Everything you read can be and has been successfully applied in those environments.

What's Ahead?

"SPACES," is an acronym for Sensor Precondition/Postcondition Assertion Check of Environmental Situations. In Chapter 9, we will discuss how SPACES can be used to verify that it's okay for the robot to carry out its autonomous tasks.

ROBOT SPACES

Robot Sensitivity Training Lesson #9: *If you're not intimate with the robot's programming, then don't invade its space.*

In Chapter 8, "Getting Started with Autonomy: Building Your Robot's Softbot Counterpart," we programmed our robot (Unit1) to autonomously approach an object, determine its color, and then report. Unit1's scenario was simple and its role was simple. One of the primary approaches to autonomous robot programming is to keep the robot's tasks well defined, the scenario and situations as simple as possible, and the robot's physical environment predictable and under control.

There are approaches to robotics that attempt to program a robot to deal with unknown, uncontrolled, and unpredictable environments, surprises, and impromptu tasks. But, so far most of these approaches require some sort of remote-control, or teleoperation. Further, the nature of the tasks that the robot can perform under these conditions is limited for many reasons (chief among them safety).

At best this approach to robotics is advanced and not recommended for beginners. Our approach to robot programming relies on well-defined tasks, scenarios, situations, and controlled environments. This allows the robot to be autonomous. Under these circumstances the robot's limits are dictated by its hardware capabilities and the skill of the robot programmer.

However, even when a robot's task and situation are well defined, things can go wrong. A robot's sensors may malfunction, motors and actuators may slip. The robot's batteries and power supplies may run down. The environment may not be exactly as expected. In Chapter 4, "Checking the Actual Capabilities of Your Robot," we discussed the process of discovering a robot's basic capabilities. Recall that even if a robot's sensors, actuators, and end-effectors are working there is always a limit to their precision. A good approach to autonomous programming is to build these considerations into the robot's programming, tasks, and role. If we don't, we cannot reasonably expect the robot to complete its tasks.

For example, what if the object from Unit1's scenario in Chapter 8 had a color that was outside the range (mauve or perhaps fuchsia) of the 16 colors that Unit1 could recognize? What would happen in this case? What if there was some obstacle between Unit1 and the object preventing Unit1 from getting close enough to determine the color, or preventing Unit1 from being able to scan the color even if it is close enough? Of course, we should know what the limitations of Unit1's color sensor are, and we should not expect it to detect the color of an object that falls outside the range.

Further, in a well-defined environment, we should be aware of any potential obstacles between Unit1 and its target. But what if Unit1's actuators slip a little, and it only travels 15 cm when it was supposed to travel 20 cm? This could throw off our plans and programming for Unit1. Since no situation or scenario can be completely controlled and the unanticipated is inevitable, what do we do when things don't turn out exactly as planned? We use an approach to programming expectation that we refer to as SPACES.

A Robot Needs Its SPACES

SPACES is an acronym for Sensor Precondition/Postcondition Assertion Check of Environmental Situations. We use robot SPACES to verify whether it is okay for the robot to carry out its current and next task. Robot SPACES is an important part of our approach to programming a robot to execute its tasks autonomously. If a robot's SPACES has been violated, corrupted, or unconfirmed, the robot is directed to report the SPACES violation and to stop task execution. If the robot's SPACES check out, then it means it is okay for the robot to execute its current task and possibly begin the execution of its next task.

The Extended Robot Scenario

In the robot scenario from Chapter 8, Unit1 is located in a small room containing a single object. Unit1 is playing the role of an investigator and is assigned the task of locating the object, determining its color, and reporting that color. In the extended scenario, Unit1 has the additional task of retrieving the object and returning it to the robot's initial position. An RSVP is a visual plan of the robot's scenario/situation and its task.

Recall from Chapter 3, "RSVP: Robot Scenario Visual Planning," that an RSVP contains the following:

- A diagram of the physical layout of the robot's scenario

- A flowchart of the robot's execution routine(s)

- A statechart showing the situation transitions that take place in the scenario

Figure 9.1 is a layout of Unit1's scenario. It shows the size of the area, the location of Unit1 within the area, and the location of the object Unit1 is to approach.

The visual layout of the robot's scenario should specify the proper shapes, sizes, distances, weights, and materials of anything that the robot has to interact with in its scenario, and anything that is relevant to the robot's role in the scenario.

For example, the layout of Unit1's physical area is 200 cm × 300 cm. The object that Unit1 has to approach is a sphere that weighs 75 grams and has a circumference of approximately 18 cm. The coordinate (0,0) is located in the southwest corner of the area and represents Unit1's starting position and final position.

Figure 9.1
The layout of *Unit1*'s scenario.

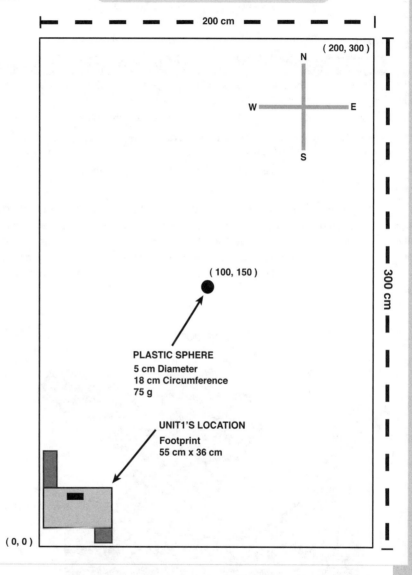

FLOORPLAN OF UNIT1'S SCENARIO

To start with, we use a simple two-dimensional coordinate system to describe the robot's location relative to the object's location. The robot is located at coordinate (0,0), and the object is located at coordinate (100,150), approximately in the center of the area. Why all the detail? Why do we need to specify locations, weights, sizes, distances, and so on? If this were a remote-controlled robot, the operator would effectively be the eyes and ears of the robot, and some of this detail would not be necessary. The operators would then use common sense and experience while controlling the robot with the remote control. But the robot does not have its own common sense and experience (at least not yet). Since we are programming our robot to be autonomous, the robot needs its own information and knowledge to be effective. Table 9.1 lists the four basic ways for an autonomous robot to obtain information and knowledge about its scenario.

Table 9.1 Four Basic Ways Autonomous Robots Obtain Information for a Scenario

Number	Method
1	The information and knowledge are explicitly stored in the robot through the programming process.
2	The robot uses its sensors to get information about its scenario and environment.
3	The robot uses inference procedures to "figure out" information about its scenario and environment.
4	Various combinations of methods 1 through 3.

Here we use a combination of methods 1 and 2 from Table 9.1 to enable the robot to have enough information to perform its task. Both methods require detail of the physical aspects of the robot's scenario and environment. Part of the information and knowledge is explicitly given to Unit1 through programming, and part the robot experiences through the use of its sensors. For this scenario, Unit1 is only equipped with an ultrasonic range finder sensor, a color sensor, a method of movement (in this case tractors), and a robot arm. Figure 9.2 is a photo of Unit1.

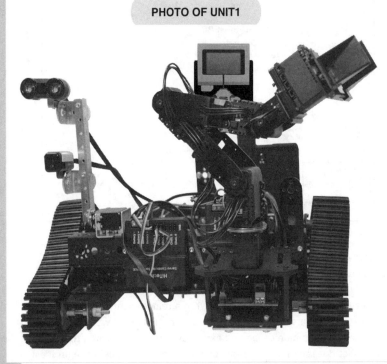

PHOTO OF UNIT1

Figure 9.2
A photo of Unit1.

The REQUIRE Checklist

Once you know what scenario the robot will be in and what role the robot will play, the next step is to determine whether the robot is actually capable of executing the task. The REQUIRE checklist can be used for this purpose. Table 9.2 shows a simple REQUIRE checklist.

 note

Remember that REQUIRE stands for Robot Effectiveness Quotient Used in Real Environments.

Table 9.2   REQUIRE Checklist for the Extended Scenario

Robot's Effectiveness	Yes/No
Sensor capable?	Yes
End-effectors capable?	Yes
Actuators capable?	Yes
Microcontroller capable?	Yes

In our extended scenario, the robot has to identify the color of an object. Since our robot does have a color sensor, we assume the robot sensors are up to the task. The robot has to retrieve a plastic sphere that weighs about 75 grams. Can Unit1 do this? Table 9.3 is a simple capability matrix for Unit1.

Table 9.3   A Simple Capability Matrix for Unit1

Robot Name	Micro-controller/ Controllers	Sensors	End-Effectors	Mobility	Communications
Unit1	ARM9 (Java): • Linux OS • 300 MHz • 16 MB Flash • 64 MB RAM 1 HiTechnic Servo Controller 1 HiTechnic DC Controller	Sensor Array: • Color Light • Ultrasonic Touch (Gripper) Smartphone Camera	Front Right Arm - 6 DOF PhantomX Pincher w/Linear Gripper Back Left Arm - 2 DOF w/ Angular Gripper	Tractor Wheeled	USB Port Bluetooth

The capability matrix specifies that Unit1's robot arm can lift about 500 grams. So 75 grams should be no problem. Unit1 has tractors and can move around the designated area, so we have a "yes" for the actuator check. Unit1 has two microcontrollers; a Mindstorms EV3 controller and an Arduino

Uno controller (for the robot arm) are also both up to their tasks. Recall that the robot's overall potential effectiveness can be measured by using a simple REQUIRE checklist.

- Sensors — 25%
- End-Effectors — 25%
- Actuators — 25%
- Controllers — 25%

Since we have a "yes" in every column on the REQUIRE checklist, we know before the robot tries to execute the task that it has a 100% potential to execute the tasks. However, the potential to execute the tasks and actually executing the tasks are not always the same thing. It is useful to look at the four REQUIRE indicators after the robot has attempted to execute the tasks to see how it performed in each area. Maybe sensors need to be changed or sensor programming needs to be tweaked. Perhaps the robot arm does not have the right DOF (Degrees of Freedom), or could lift 75 grams but not hold 75 grams long enough.

The REQUIRE checklist can be used in a before/after manner to determine whether the robot has the potential to perform the tasks and how well the robot actually performs the task. If the robot cannot pass the checklist initially, the indicators have to be adjusted, changed, or improved until the checklist is passed. Otherwise, we know the robot will not be capable of effectively performing the tasks. If the robot passes the checklist, and you have a diagram of the physical layout of the robot's scenario, the next step is to produce a diagram of the procedure that shows step-by-step what actions the robot needs to execute to successfully perform its role in the scenario. This is an important part of the process for several reasons:

- It helps you figure out what the robot is doing prior to making the effort to write the code.
- It helps you discover aspects or details about the scenario that you had not thought about.
- You can use this diagram as a reference for future use for design and documentation purposes.
- You can use it as a communication tool when sharing the idea of the robot's task with others.
- It helps you find mistakes or errors in your robot's logic.

Figure 9.3 shows a simplified version of a flowchart for `Unit1`'s extended scenario actions. We marked eight of the actions that the robot has to execute as steps 1 through 8. We use these process boxes as a shortcut here. For example, the robot initialization in step 1 represents several steps such as initializing the sensors and setting the sensors into the proper modes, for example, analog or digital, initializing the motors and setting the initial speed for the motors, and positioning the robot arm into its initial position.

 note

Note that the boxes for step 1 and step 3 have double lines. Remember from Chapter 3 that double lines mean that those boxes represent multiple steps or a subroutine that can be broken down into simpler steps.

Figure 9.3
A simplified version of
Unit1's extended scenario
actions flowchart

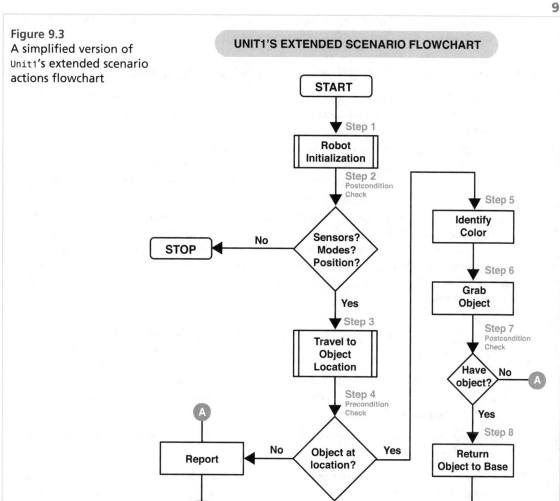

UNIT1'S EXTENDED SCENARIO FLOWCHART

Robot initialization includes the robot checking to see whether it has enough power supply to complete the task. The initialization also includes the robot checking its initial location. Hopefully the robot is starting out in the right place to begin with. We could have separate boxes for each of these steps, but for now, to keep things simple we use a single process box for initialization in step 1 and the robot travel in step 3.

The initialization done in step 1 is usually done in some sort of setup() or startup() routine, and in all the examples in this book those routines are part of a Java or C++ (Arduino) constructor. In step 2, step 4, and step 7 from Figure 9.4, we check Unit1's SPACES. That is, we check the preconditions, postconditions, and assertions about the robot's situation. A *precondition* is some condition that must be true at the start of, or prior to, some action taking place. A *postcondition* is a condition that should be true immediately after an action has taken place.

What Happens If Pre/Postconditions Are Not Met?

If the preconditions for a robot's action have not been met or are not true, what should we do? If the postconditions for a robot's action are not true, what does that mean? Should the robot continue to try to perform the action if the action's preconditions are not true?

For instance, in step 2 from Figure 9.3, there is a postcondition check for the robot's initialization process. The postcondition checks to see whether the sensors have been put in the proper modes and if the motors have been set to the proper speeds. The postcondition checks to see whether the robot's arm is in the correct starting position and so on.

If these conditions haven't been met, should we send the robot on its mission anyway? Is that safe for the robot? Is that prudent for the situation? Notice that our flowchart says if the postconditions in step 2 are not met, we want the robot to stop. If the postconditions are not met, for all we know the robot is undergoing a significant malfunction and we don't want the robot to even attempt to execute the task. When the pre/postconditions or assertions are not met or are not true, we call this *violating the robot's SPACES*.

What Action Choices Do I Have If Pre/Postconditions Are Not Met?

What are the available choices for the robot's next action if some precondition or postcondition is not met? We look at three basic actions the robot could take. Although there are more, they are essentially variations on these three:

1. The robot could ignore the pre/postcondition violation and continue trying to execute any action that it can still execute.

2. The robot could attempt to fix or rectify the situation in some way by retrying an action or adjusting some parameter or the position of one or more of its parts, actuators, or end-effectors. This would amount to the robot trying to make the pre/postcondition true if it can and then proceed with the task.

3. The robot can report the nature of the pre/postcondition violation, put itself into a safe state, and then stop or shut down.

Much of the challenge of robot programming is related to the problem of what to do next if something does not go as planned or expected. Entire approaches to robot programming have been developed around the problems of unknown environments, surprises within the environment, a changing or evolving situation, and robot malfunction. But to get us started with programming a robot to be autonomous, we program for situations that are under our complete control, where the environment is well defined, and where we can anticipate, test for, and react to preconditions and postconditions that are met or not met. We start with closed situations and scenarios and learn how to give a robot level 1 through level 3 autonomy (refer to Table 9.1).

In Chapter 12, "Open Source SARAA Robots for All!," we explain SARAA (**S**afe **A**utonomous **R**obot **A**pplication **A**rchitecture) in comparison with pure reactive approaches to robot autonomy.

A Closer Look at Robot Initialization Postconditions

The robot's initialization routine is executed in step 1 in the flowchart shown in Figure 9.3. Since step 1 uses a process box, there are multiple steps or a subroutine of steps involved. We use our BURT translator to show the list of actions taken in the initialization routine. Listing 9.1 shows the BURT translation for Unit1's constructor (i.e., initialization routine) from step 1 in Figure 9.3.

Listing 9.1   Burt Translation Initialization Routine/Constructor

BURT Translator INPUT

```
Softbot  Frame
Name:  Unit1


Initialization Routine Start:
Step 1: Initialize ultrasonic sensor
Step 2: Initialize color sensor
Step 3: Setup left and right motors
Step 4: Initialize arm servos
        Set arm to initial angle of 100
Step 5: Set robot's initial position to 0,0
Step 6: Set wheel diameter to  7
        Set track width to 32
        ...
If robot's startup routine is successful then continue, otherwise report problems,
and shutdown.
Initialization Routine  End:
```

BURT Translator OUTPUT

```
{
    1   public unit1() throws InterruptedException,Exception
    2   {
    3
    4       try{
    5           Exception SoftbotError;
    6           //Set up ultrasonic sensor
    7           Vision = new UltrasonicSensor(SensorPort.S4);
    8           if(Vision == null){
```

```
 9          Messages.add("Could Not Initialize Ultrasonic Sensor on Port 4");
10          SoftbotError = new Exception("101");
11          throw SoftbotError;
12      }
13      Vision.enable();
14      //Set up color sensor
15      ColorVision = new HiTechnicColorSensor(SensorPort.S2);
16      if(ColorVision == null){
17          Messages.add("Could Not Initialize Color Sensor on Port 2");
18          SoftbotError = new Exception("100");
19          throw SoftbotError;
20      }
21      // set up motors
22      CF = new TetrixControllerFactory(SensorPort.S1);
23      if(CF == null){
24          Messages.add("Could Not Setup Servo Port");
25          SoftbotError = new Exception("102");
26          throw SoftbotError;
27      }
28
29      LeftMotor = MC.getRegulatedMotor(TetrixMotorController.MOTOR_1);
30      RightMotor = MC.getRegulatedMotor(TetrixMotorController.MOTOR_2);
31      if(LeftMotor == null || RightMotor == null){
32          Messages.add("Could Not Initialize Motors");
33          SoftbotError = new Exception("103");
34          throw SoftbotError;
35      }
36
37      LeftMotor.setReverse(true);
38      RightMotor.setReverse(false);
39      LeftMotor.resetTachoCount();
40      RightMotor.resetTachoCount();
41
42      //Set up arm servos
43      Messages.add("Tetrix Controller Factor Constructed");
44      MC = CF.newMotorController();
45      SC = CF.newServoController();
46      Gripper = SC.getServo(TetrixServoController.SERVO_2);
47
48
49      Arm = SC.getServo(TetrixServoController.SERVO_1);
50      if(Arm == null){
51          Messages.add("Could Not Initialize Arm");
52          SoftbotError = new Exception("104");
53          throw SoftbotError;
54      }
55      Arm.setRange(750,2250,180);
```

```
56              Arm.setAngle(100);
57              // Set Robot' initial Position
58              RobotLocation = new location();
59              RobotLocation.X = 0;
60              RobotLocation.Y = 0;
61
62
63              //Set  Wheel Diameter and Track
64              WheelDiameter = 7.0f;
65              TrackWidth = 32.0f;
66
67              Situation1 = new situation();  // creates new situation
68
69
70      }
71  }
//Burt Translation End Constructor
```

Power Up Preconditions and Postconditions

The first preconditions and postconditions are usually encountered in the constructor. Recall the constructor is the first code that gets executed whenever an object is created. When the softbot (control code) for Unit1 starts up, the first thing executed is the constructor. The initialization, startup, or power up routines are especially important for an autonomous robot. If something goes wrong in the startup sequence, all bets are off. The robot's future actions are not reliable if the power up sequence fails in some way.

In the BURT translator for Listing 9.1, we show six of Unit1's startup routine steps. Keep in mind that at this stage of the RSVP, when we are specifying the design of the steps the robot is to take, we want to express each step simply and make it easy to understand. It's important for the list of instructions to be clear, complete, and correct. Once you understand the instructions you will give the robot in your language, it's time to translate those instructions into the robot's language. The output translation in Listing 9.1 shows what the steps will look like once they are translated into Java. Ideally we name the variables, routines, and methods so that they match the input design language as closely as practical.

This is our first postcondition. We call it a postcondition because we check whether it's true *after* some list of actions has been attempted or executed. In this case, the actions are steps 1 through 6. Our initial policy is that it's better to be safe than sorry. If any of the actions fail in steps 1 through 6, we do not send the robot on its mission. For example, if the ultrasonic sensor could not attach to port 3, or if we could not set the left and right motors, this would be a SPACES violation because one of the postconditions of our constructor requires that the start routines have to be successful. So how do we code preconditions and postconditions?

 note

Notice the rule in the translator's input: *If a robot's startup routine is successful then continue; otherwise report problems and shut down.*

Coding Preconditions and Postconditions

Let's look at lines 5 through 12 from the BURT translation in Listing 9.1:

```
5          Exception SoftbotError;
6          //Set up ultrasonic sensor
7          Vision = new UltrasonicSensor(SensorPort.S4);
8          if(Vision == null){
9              Messages.add("Could Not Initialize Ultrasonic Sensor on Port 4");
10             SoftbotError = new Exception("101");
11             throw SoftbotError;
12         }
```

We instruct the robot to initialize the ultrasonic sensor on line 7. Lines 8 through 12 determine what happens if that action could not be performed. There is an instruction to initialize on line 7 and a condition to check afterward to see whether that action was taken. This is what makes it a postcondition. First the action is tried; then the condition is checked. There are some other interesting actions taken if the ultrasonic sensor is not initialized, that is, (Vision == null). We add the message

```
"Could Not Initialize Ultrasonic Sensor on Port 4"
```

to the Messages ArrayList. The Messages ArrayList is used to log all the important robot actions and actions that the robot failed to execute. This ArrayList is later either saved for future inspection or transmitted over a serial, Bluetooth, or network connection to a computer so that it can be viewed. After a message is added, we create a SoftbotError Exception("101") exception object and then we throw this object. Any code after line 11 in the constructor is not executed. Instead, robot control is passed to the first exception handler that can catch objects of type exception and the robot comes to a halt.

Notice lines 11, 19, 26, and 34 all throw an Exception object. If any one of these lines is executed, the robot comes to a halt and does not proceed any further. If any one of those lines is executed it means that a postcondition was not met, and the first missed postcondition ultimately causes the robot to come to a halt. Notice in Listing 9.1 earlier in the chapter that the constructor has five postcondition checks:

```
// Postcondition 1
8    if(Vision == null){
     ...
12   }


// Postcondition 2
16   if(ColorVision == null){
     ...
20   }


// Postcondition 3
23   if(CF == null){
     ...
```

```
27     }
```

```
// Postcondition 4
31     if(LeftMotor == null || RightMotor == null){
       ...
35     }
```

```
// Postcondition 5
50     if(Arm == null){
       ...
54     }
```

What is the effect of these five postcondition checks in the constructor? If there are any problems with the robot's vision, color vision, servos, motors, or arm, the robot's mission is cancelled plain and simple. In this case, the checks are made using if-then control structures. Recall the control structures introduced in Chapter 3. We could use any of five control structures shown in Figure 9.4 to check preconditions/postconditions and assertions.

Structure 1 in Figure 9.4 says if a condition is true, then the robot can take some action. If Structure 1 is used to check a condition after a set of actions have been performed, then Structure 1 is being used to check a postcondition.

If Structure 1 is being used to check a condition before one or more actions will be taken, then Structure 1 is being used to check a precondition. Where Structure 1 is a one-time check, Structures 2 and 3 perform one or more actions while a condition is true (Structure 2) or until a condition becomes true (Structure 3). Structure 4 is used when a condition is to be selected from a group of conditions and a separate action(s) needs to be taken depending on which condition is true. Structure 5 is used to handle abnormal, unanticipated conditions. What is important is that we decide which conditions if not met are showstoppers and which are not. The setup routine, startup routine, initialization routine, and constructor are the first place pre/postconditions should be used. If the robot does not power up or start out successfully, usually it's downhill from there. There are exceptions. We could include recovery and resumption routines. We could endow our robot with fault tolerance and redundancy routines. We introduce some of these techniques later, but for now, we stick with the basics. The robot is only to proceed if the power up sequence is successful; otherwise, "abort mission"!

In step 3 of the flowchart from Figure 9.3 earlier in the chapter, the robot is given the instruction to travel to the object's location. On lines 58 through 60 in Listing 9.1, we set the robot's initial location. For the robot to travel to the object's location, the robot needs to know where it is starting from and where to travel to. This information is part of the robot's situation and scenario. The initial X and Y location equals 0 and is a precondition of the robot's travel action. If the robot is not at the proper location to start with, then any further directions that it follows will not be correct. Keep in mind that we are programming our robot to be autonomous. It will not be directed by remote control to the correct location. The robot will be following a set of preprogrammed directions. If the robot's starting position is incorrect (i.e., the precondition is not met), then the robot will not successfully reach its destination.

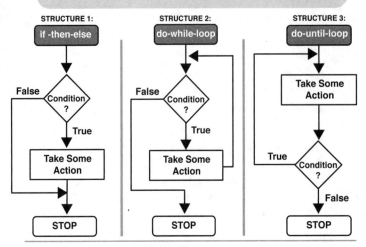

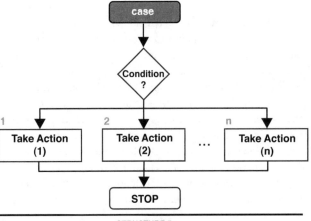

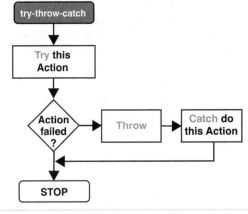

Figure 9.4
Basic control structures to check preconditions and postconditions

Once the robot gets to the object's location, the robot must determine the object's color. Is there a precondition here? Yes! The robot cannot determine the object's color if the object is not at the designated location. So the precondition is the object must be at the designated location. Step 4 in Figure 9.4 performs this precondition check. Let's take a look at the actual programming of steps 3 and 4 to see what this looks like. In step 3, the robot is given the instruction to travel. The method of travel differs depending on the robot's actuators and whether the robot's build is

- Bipedal

- Quadruped/hexaped, and so on

- Tractors/wheels

- Underwater

At the lowest levels (level 1 and level 2 of the ROLL) of programming, the robot travelling (or moving) is accomplished by directly programming the motors using microcontroller commands directed to motor/servo ports and pins.

 note

See Chapter 7, "Programming Motors and Servos," for a closer look at programming robot motors.

Activating motors and servos is what causes a robot's movement. Programming the motors to rotate in one direction causes wheels, legs, propellers, and so on to move in that direction. The accuracy of the movement depends on how much precision the motor has and how much control you have over the motor. Unregulated or DC motors may be appropriate for certain kinds of movement, and step-controlled motors may be more appropriate for others.

Low-level motor programming can be used to translate motor rotations or steps into distance. However, even low-cost robot environments, such as RS Media, Arduino, and Mindstorms EV3 and NXT, have motor/movement class libraries that already handle a lot of the low-level detail of programming motors and servos to move. Table 9.4 shows examples of some of the commonly used class libraries for robot motor control.

Table 9.4 Examples of Commonly Used Motor and Servo Classes for Low-Cost Robots

Class/Library	Language	Robot Environment
Servo	C++	Arduino
BasicMotor	Java (leJOS)	NXT
TetrixRegulatedMotor	Java (leJOS)	NXT
EV3LargeRegulatedMotor	Java (leJOS)	EV3
Ev3MediumRegulatedMotor	Java (leJOS)	EV3
Servo	Java	RS Media
Walk	Java	RS Media

At higher robot programming levels (for example, levels 3 through 5 of the ROLL), programming the robot to travel() involves using classes such as the ones shown in Table 9.4 and calling methods or functions provided by those classes. For example, the Arduino environment has a Servo class. The Servo class has a write() method. If we wanted to make an Arduino servo move 90 degrees, we could code the following:

```
#include <Servo.h>

Servo  Servo1;  //Create an object of type Servo called Servo1
int  Angle = 90;
void setup()
{
    Servo1.attach(9)    //Attach the servo on pin 9
    if (Servo1.attached()){
        Servo1.write(Angle);
    }
}
```

This gets the servo to move 90 degrees. And if the servo is connected to the robot's wheels, tractors, legs, and so on, it causes some kind of 90-degree movement. But what does this type of program have to do with traveling or walking? General travel(), walk(), and move() procedures can be built based on methods and functions that classes like the Arduino Servo or the leJOS TetrixRegulatedMotor provides. So you would build your routines on top of the built-in class methods. Before using a class, it is always a good idea to familiarize yourself with the class's methods and basic functionality. For instance, some of the commonly used member functions (or methods) for the Arduino Servo class are shown in Table 9.5.

Table 9.5 Commonly Used Methods of the Arduino Servo Class

Method	Description
attach(int)	Turn a pin into a servo driver. Calls pinMode.
detach	Release a pin from servo driving.
write(int)	Set the angle of the servo in degrees.
read()	Return that value set with the last write( ).
attached()	Return 1 if the servo is currently attached( ).

If your robot is intended to be mobile, you should include some kind of generic travel(), moveForward(), moveBackward(), reverse(), and stop() routines that are built on top of the class methods provided by the library for your microcontroller. Routines involving sensor measurements or motor or servo movement ultimately use some unit of measure. Our travel routine from step 3 in Figure 9.3 has to assume some unit of measure. Will the robot's moveForward(), travel(), or reverse() routines use kilometers, meters, centimeters, and so on? When you program a robot to move, you should have specific units of measure in mind. In our programming examples, we use centimeters. In Figure 9.3, steps 3 through step 5, we program the robot to travel to the object, check to see whether the object is there, and take some action. The BURT translation in Listing 9.2 shows our level 5 instructions and their level 3 Java implementations.

Listing 9.2   BURT Translation Traveling to Object

BURT Translator INPUT

```
Softbot  Frame
Name:  Unit1 Level 5

Travel to the object Algorithm Start:
if the object is there  {this is precondition}
determine its color.
Travel to the object Algorithm End.
```

BURT Translator OUTPUT: Level 3

```
//Begin Translation

1   Unit1 = new softbot();
2   Unit1.moveToObject();
    ...
5   Thread.sleep(2000);
6   Distance = Unit1.readUltrasonicSensor();
7   Thread.sleep(4000);
8   if(Distance <= 10.0){
9      Unit1.getColor();
10     Thread.sleep(3000);
11  }

//Translation End.
```

The code in Listing 9.2 directs the robot to `moveToObject()` and then take a reading with the ultrasonic sensor. Keep in mind that ultrasonic sensors measure distance. The assumption here is that the robot will move within 10 cm of the object. Notice on Line 8 that we check to see whether `Distance` from the object is `<= 10.0`. This is the precondition that we referred to in step 4 of Figure 9.4. If the precondition has not been met, what do we do?

If the robot is at the correct location and there is no object within 10 cm, then how can it determine the object's color? If there is an object located 12, 15, or 20 centimeters from the robot, how do we know that is the object we want to measure? In our case, we specify an object that is 10 cm or less from the robot's stopping position. If there is no object there, we want the robot to stop and report the problem. It's not a requirement that the robot always stop when its SPACES have been violated. However, it is important to have some consistent well-thought-out plan of action for the robot if its SPACES are violated during the execution of its program.

Where Do the Pre/Postconditions Come From?

How do we know the robot should stop within 10 cm of the object? When we give the robot the command to `moveToObject()` in Listing 9.2, what object are we referring to? Where is it? Recall that SPACES is an acronym for Sensor Precondition/Postcondition Assertion Check of Environmental

Situations. *Situations* is a keyword in this acronym. In our approach to programming robots to be autonomous, we require the robot to be programmed for specific scenarios and situations. The situation is given to the robot as part of its programming. The pre/postconditions are a natural part of the situation or scenario that has been given to the robot. That is, the situation or scenario dictates what the preconditions or postconditions are. Recall the high-level overview of our robot's situation from Chapter 8:

> The Extended Robot Scenario
>
> Unit1 is located in a small room containing a single object. Unit1 is playing the role of an investigator and is assigned the task of locating the object, determining its color, and reporting that color. In the extended scenario, Unit1 has the additional task of retrieving the object and returning it to the robot's initial position.

By the time this high-level situation is translated into details the robot can act upon, a lot of questions are answered. For example, the extended robot scenario makes several statements that immediately pose certain questions:

- **Statement:** Unit1 is located in a small room.

- **Question(s):** Where in the room?

- **Statement:** ...containing a single object.

- **Question(s):** Where is the object? How big is it?

- **Statement:** ...returning it to the robot's initial position.

- **Question(s):** Where is the initial position? Where should the object be placed?

These kinds of statements and questions help to make the preconditions and postconditions. Let's take a look at our Java implementation of Unit1.moveToObject() shown in Listing 9.3.

Listing 9.3   BURT Translation of moveToObject() method

BURT Translator OUTPUT: Java Implementation

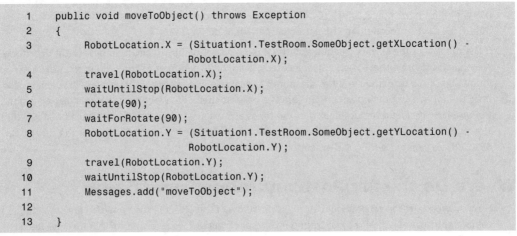

```
1    public void moveToObject() throws Exception
2    {
3        RobotLocation.X = (Situation1.TestRoom.SomeObject.getXLocation() -
                            RobotLocation.X);
4        travel(RobotLocation.X);
5        waitUntilStop(RobotLocation.X);
6        rotate(90);
7        waitForRotate(90);
8        RobotLocation.Y = (Situation1.TestRoom.SomeObject.getYLocation() -
                            RobotLocation.Y);
9        travel(RobotLocation.Y);
10       waitUntilStop(RobotLocation.Y);
11       Messages.add("moveToObject");
12
13   }
```

The robot has an X and a Y location. The robot's X location is determined on line 3 in Listing 9.3, and the robot's Y location is determined on line 8 of Listing 9.3. The instruction

```
RobotLocation.X = (Situation1.TestRoom.SomeObject.getXLocation() - RobotLocation.X);
```

subtracts the robot's current X position from the location of the object in the test room. This gives the distance in centimeters the robot is to travel east or west. If RobotLocation.X is positive after the subtraction, the robot travels west, and if RobotLocation.X is negative, the robot travels east. A similar calculation is used to determine how far the robot travels north or south with the instruction:

```
RobotLocation.Y = (Situation1.TestRoom.SomeObject.getYLocation() - RobotLocation.Y);
```

If RobotLocation.Y is positive after the subtraction, the robot travels north, or the robot travels south if RobotLocation.Y is negative after the subtraction. But notice the following object construction:

```
Situation1.TestRoom.SomeObject
```

This means there is an object called Situtation1 that has a component named TestRoom, and TestRoom has a component named SomeObject. The component SomeObject has a getXLocation() and getYLocation() method that returns the X,Y coordinates (in centimeters) of SomeObject. We look closer at the technique of programming the robot with one or more situations in Chapter 10, "An Autonomous Robot Needs STORIES." Here, we show in Listing 9.4 the class declarations for situation, x_location, room, and something.

Listing 9.4 BURT Translations for situation, x_location, room, **and** something **Classes**

BURT Translations Output: Java Implementations

```
1    class x_location{
2        public int X;
3        public int Y;
4        public x_location()
5        {
6
7            X = 0;
8            Y = 0;
9        }
10
11    }
12
13
14   class something{
15       x_location Location;
16       int Color;
17       public something()
```

```
18        {
19            Location = new x_location();
20            Location.X = 0;
21            Location.Y = 0;
22            Color = 0;
23        }
24        public void setLocation(int X,int Y)
25        {
26
27            Location.X = X;
28            Location.Y = Y;
29
30        }
31        public int getXLocation()
32        {
33            return(Location.X);
34        }
35
36        public int getYLocation()
37        {
38            return(Location.Y);
39
40        }
41
42        public void setColor(int X)
43        {
44
45            Color = X;
46        }
47
48        public int getColor()
49        {
50            return(Color);
51        }
52
53    }
54
55    class room{
56        protected int Length = 300;
57        protected int Width = 200;
58        protected int Area;
59        public something SomeObject;
60
61        public  room()
62        {
63            SomeObject =  new something();
64            SomeObject.setLocation(20,50);
```

```
65              }
66
67
68          public int  area()
69          {
70              Area = Length * Width;
71              return(Area);
72          }
73
74          public  int length()
75          {
76
77              return(Length);
78          }
79
80          public int width()
81          {
82
83              return(Width);
84          }
85      }
86
87      class situation{
88
89          public room TestRoom;
90          public situation()
91          {
92              TestRoom = new room();
93
94          }
95      }
```

With Listing 9.4 in mind, we declare

```
situation  Situation1;
```

to be a component of Unit1's softbot frame. Classes like room, situation, something, and x_
location provide the basis for the pre/postconditions that make up our robot's SPACES. If we look
at line 64 in Listing 9.4, we see that the object is located at (20,50). If the precondition

```
if(Distance <= 10.0)
```

from line 8 in Listing 9.2 is satisfied, this means that once the robot has executed its
moveToObject() instruction it should be <= 10 centimeters from the location (20,50).

SPACES Checks and RSVP State Diagrams

Recall the introduction to state diagrams in Chapter 3. We use each state to represent a situation in the scenario. So we can say a scenario consists of a set of situations. A good rule of thumb is to have one or more pre/postcondition checks before every situation the robot can be in during the scenario. If the SPACES are not violated, it is safe or okay for the robot to proceed from its current situation to the next situation. Figure 9.5 shows the seven situations the object color recognition and retrieval scenario has.

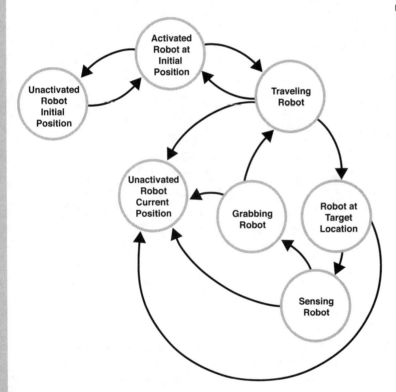

SITUATION STATE DIAGRAM OF ROBOT'S SCENARIO

Figure 9.5
A state diagram that shows the seven situations of the robot's scenario.

Drawing the type of diagram in Figure 9.5 helps you plan your robot's autonomy. It helps to identify and clarify what situations the robot will be in during the scenario. This kind of diagram helps you see where the robot's SPACES are. Typically, at least one pre/postcondition needs to be met in every situation. For autonomous robots, there is usually more than one pre/postcondition per situation. Let's revisit the three basic tools of the RSVP we used for our extended robot scenario. The first tool is a physical layout of the robot's scenario from Figure 9.1.

This layout gives us some idea where the robot starts out, where it has to go, where the object is located, what size the area is, and so on. At this level the layout is a snapshot of the robot's initial situation within the scenario. The second tool is the flowchart of the set of instructions the robot is to carry out from Figure 9.3. This tool shows the primary actions the robot is to perform autonomously and where the preconditions and postconditions occur within the robot's set of instructions. This tool gives us a picture of the planned actions and decisions the robot must execute autonomously. The final tool is the situation/scenario state diagram from Figure 9.5 that shows how the scenario is broken down into a set of situations and how the robot can legally move between situations. For example, in Figure 9.5, the robot cannot go from a traveling situation directly to a grabbing situation, or from a grabbing situation to a sensing situation. The situation/scenario statechart shows us how the situations are connected in the scenario and where the SPACES are likely to occur.

Although the RSVP is used as a set of graphic plans that support the identification and planning of the robot's SPACES and REQUIRE checklist, keep in mind that components depicted in those three diagrams must have C++ or Java counterparts (depending on which language you are using for implementation). Listing 9.4 contains the Java classes used to implement the situations of our robot's extended scenario, and the BURT translations in Listing 9.1 and Listing 9.2 are examples of how some of the robot's SPACES are implemented in Java. We showed how SPACES can be used to identify the particular C++ or Java class components and methods and Arduino classes such as the Servo class. From this we can see that the visual tools and techniques of the RSVP are helpful both for the visualizing and the planning of our robot's autonomy.

What's Ahead?

In Chapter 10, "An Autonomous Robot Needs STORIES," we show the complete program for our robot's extended scenario. We take a closer look at SPACES and exception handling. We also introduce you to robot STORIES, the final piece of the puzzle to programming a robot to execute tasks autonomously.

AN AUTONOMOUS ROBOT NEEDS STORIES

Robot Sensitivity Training Lesson #10: *A robot doesn't know what it doesn't know.*

At the hardware level a robot is simply a combination of chips, wires, pins, actuators, sensors, and end-effectors. How do we get from those components to a robot that can light the candles on a birthday cake or clean up after a party? At ROLL level 1, programming a robot is all about setting a voltage high/low, trapping signals, and reading pins.

At level 2, we graduate to making software drivers for the robot's servos and actuators. We can control gear speeds using level 1 and level 2 robot programming. But at some point we want a robot that can do useful things in a meaningful environment. Not only that, we want the robot to do useful things without supervision and hand holding. We want the robot to walk our dog, bring us a cool beverage, turn off the lights, and close the door all on its own. All of that is a long way from setting pin voltages and stepping motors. How do we get from sending signals to a collection of hardware components to autonomous useful action by a robot?

Recall the seven requirements for our definition of a robot from Chapter 1, "What Is a Robot, Anyway?":

1. It must be capable of sensing its external and internal environments in one or more ways through the use of its programming.

2. Its reprogrammable behavior, actions, and control are the result of executing a programmed set of instructions.

3. It must be capable of affecting, interacting with, or operating on its external environment in one or more ways through its programming.

4. It must have its own power source.

5. It must have a language suitable for the representation of discrete instructions and data as well as support for programming.

6. Once initiated, it must be capable of executing its programming without the need for external intervention.

7. It must be a nonliving machine. (Therefore, it's not animal or human.)

 note

Requirement #5 specifies that a robot must have a language that can support instructions and data. A programming language has a method of specifying actions and a method of representing data or objects. An autonomous robot must carry out actions on objects in some designated scenario and environment. The language associated with the robot provides the fundamental key. Using a programming language, we can specify a list of actions the robot must execute. But the programming language can also be used to describe the environment and objects the robot must interact with. Requirements 1, 3, and 6 then can be used to implement robot autonomy.

It's Not Just the Actions!

The actions that the robot performs are usually center stage when thinking about robot programming. But the environment in which the robot performs and the objects that the robot interacts with are at least as important as the actions the robot performs. We want the robot to perform tasks. These tasks involve the robot interacting with objects within a certain context, situation, and scenario.

The process of programming the robot requires that we not only use the programming language to represent the robot's actions, but use the programming language to represent the robot's environment, its situation, and the objects the robot must interact with. The situation and object representation requirements cause us to select object-oriented languages like C++ and Java for programming a robot to be autonomous. These languages support object-oriented and agent-oriented programming techniques. And these techniques are part of the foundation of autonomous robots. Let's revisit our birthday party robot for a moment.

Birthday Robot Take 2

Recall that in our birthday party BR-1 scenario, the BR-1 was charged with the responsibility of lighting the candles on the birthday cake and then clearing the table of paper plates and cups after the party was over. How do we specify to the BR-1 what a birthday cake is? How do we describe the candles? What does it mean to light the candles? How will the BR-1 determine when the party is over? How do we program the BR-1 to recognize paper plates and cups?

In Chapter 6, "Programming the Robot's Sensors," we explain how basic data types are used with sensors. For example, we described how `floats` and `ints` might be used to represent sensor measurements of temperature, distance, and color such as:

```
float   Temperature =  96.8;
float   Distance    =  10.2;
int     Color       =  16;
```

However, simple data types are not enough to describe a birthday cake or a birthday party scenario to a robot. What simple data type could we use to represent a candle or the birthday cake? How would we describe the idea of a party in our birthday robot scenario? A task is made up of actions and things. Robot programming must have a way to express the actions that must be performed as well as the things involved in those actions. There must be some way to describe to the robot its environment. There must be some way to pass on to the robot the sequence of events that are part of the scenario we want the robot to participate in. Our approach to programming autonomous robots requires that the robot be fully equipped with a description of a scenario, situation, episode, or script. So exactly what do we mean by scenario, situation, or script? Table 10.1 shows some common definitions that we use.

Table 10.1 Common Definitions for Scenario, Situation, Episode, and Script

Concept	Commonly Found Definitions
Scenario	A sequence of events, a description of a possible course of actions
Situation	All of the facts, objects, conditions, and events that affect someone or something
Episode	An event that is distinctive and separate but part of larger scenario
Script	A standardized sequence of events that describes some stereotypical activity (i.e., going to a birthday party)

The basic idea is to give the robot what to expect, a role to play, and one or more tasks to execute and then send it on its way. Everything in this approach is *expectation driven*. We remove or at the very least reduce any surprises for the robot during the tasks it executes. This expectation-driven approach is meant to be totally self-contained. Scenarios, situations, episodes, and scripts contain all the relevant sequences of actions, events, and objects involved with the robot's interactions. For example, if we think through our birthday party scenario, all the information we need is part of the scenario. The idea of a birthday party is common.

There is:

- A celebration of some sort

- A guest of honor (it's somebody's birthday)

- Birthday guests

- Birthday cake, or possibly birthday pie

- Ice cream, and so on

Sure, some of the details of any particular birthday party may vary, but the basic idea is the same, and once we decide what those details are (for example, cake or pie, chocolate or vanilla, number of guests, trick candles, and so on), we have a pretty complete picture of how the birthday party should unfold. So if we can just package our job for the robot as a scenario, situation, script, or episode, the hard part is over. What we need to program our birthday party robot to execute its birthday party responsibilities is a way to describe the birthday scenario using object-oriented languages like Java or C++, and then upload that scenario to the robot to be executed. At Ctest Laboratories, we developed a programming technique and data storage mechanism that we call STORIES that does just that.

Robot STORIES

STORIES is an acronym for Scenarios Translated into Ontologies Reasoning Intentions and Epistemological Situations. STORIES is the end result of converting a scenario into components that can be represented by object-oriented languages and then uploaded into a robot. Table 10.2 is a simplified overview of the five-step process.

Table 10.2 Simplified Overview of the Five-Step Process to Create Robot STORIES

Step #	Basic Steps for Creating STORIES	STORIES Component
1	Break down the scenario into a sequence of events, actions, and the list of things that make up the scenario.	Ontology
2	Describe the elements from step 1 using an object-oriented language (e.g., C++ or Java).	Ontology
3	Set up rules for how the robot should use and make decisions about the elements described in step 1 (this is the reasoning part) and write those rules in the programming language.	Reasoning
4	Describe the robot's particular task(s) that is to be executed in the scenario.	Intention
5	Connect the components from step 3 as pre/postconditions to robot execution of task(s) described in step 4.	An epistemological situation

STORIES is a technique and storage mechanism that allows us to program a robot to autonomously execute a task(s) within a particular scenario. This is made possible because the scenario becomes part of the instructions uploaded to the robot.

In this book, the actions that we want the robot to perform are written as C++ or Java code, and the scenario that we want the robot to perform in is also written as Java or C++ code. STORIES makes our anatomy of autonomous robot complete. And once you've integrated the STORIES component into your robot programming, you have the basic components required to program a robot to execute tasks autonomously. Figure 10.1 is the first of several blueprints that we present in this chapter that give us an anatomy of an autonomous robot. Each blueprint provides more detail until we have a complete and detailed blueprint.

The two primary components in Figure 10.1 are the robot skeleton (hardware) and the robot's softbot frame. At this level of detail in the blueprint, the three major components of the robot's software are its basic capabilities, its SPACES (introduced in Chapter 9, "Robot SPACES"), and its STORIES. To see how all this works for our Arduino and Mindstorms EV3-based robots, let's revisit our extended robot scenario program from Chapter 9.

Figure 10.1
Blueprint of
an anatomy
of an autono-
mous robot

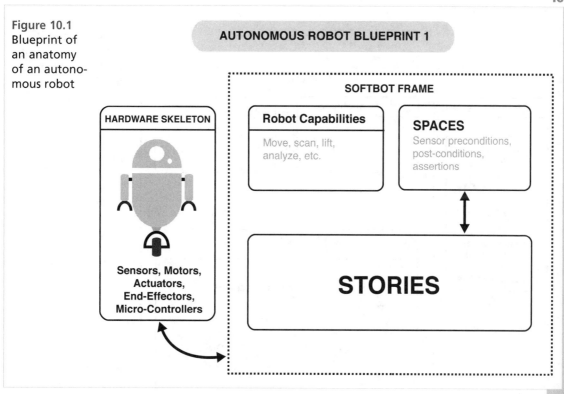

AUTONOMOUS ROBOT BLUEPRINT 1

SOFTBOT FRAME

HARDWARE SKELETON

Sensors, Motors,
Actuators,
End-Effectors,
Micro-Controllers

Robot Capabilities

Move, scan, lift,
analyze, etc.

SPACES

Sensor preconditions,
post-conditions,
assertions

STORIES

The Extended Robot Scenario

The robot (Unit1) is located in a small room containing a single object. Unit1 is playing the role of an investigator and is assigned the task of locating the object, determining its color, and reporting that color. In the extended scenario, Unit1 has the additional task of retrieving the object and returning it to the robot's initial position.

Converting Unit1's Scenario into STORIES

We use the five-step approach shown previously in Table 10.2 to convert or translate our extended robot scenario into STORIES that can be uploaded for Unit1's execution. The five steps represent each of the major parts of the STORIES software components. Figure 10.2 shows a more detailed blueprint of our autonomous robot anatomy.

First, let's break our scenario down into a list of things, actions, and events. There are many ways we can do this, and we can do it in varying levels of detail. But here we opt for a simple breakdown. Table 10.3 shows three things, three events, and three actions that make up our extended robot scenario.

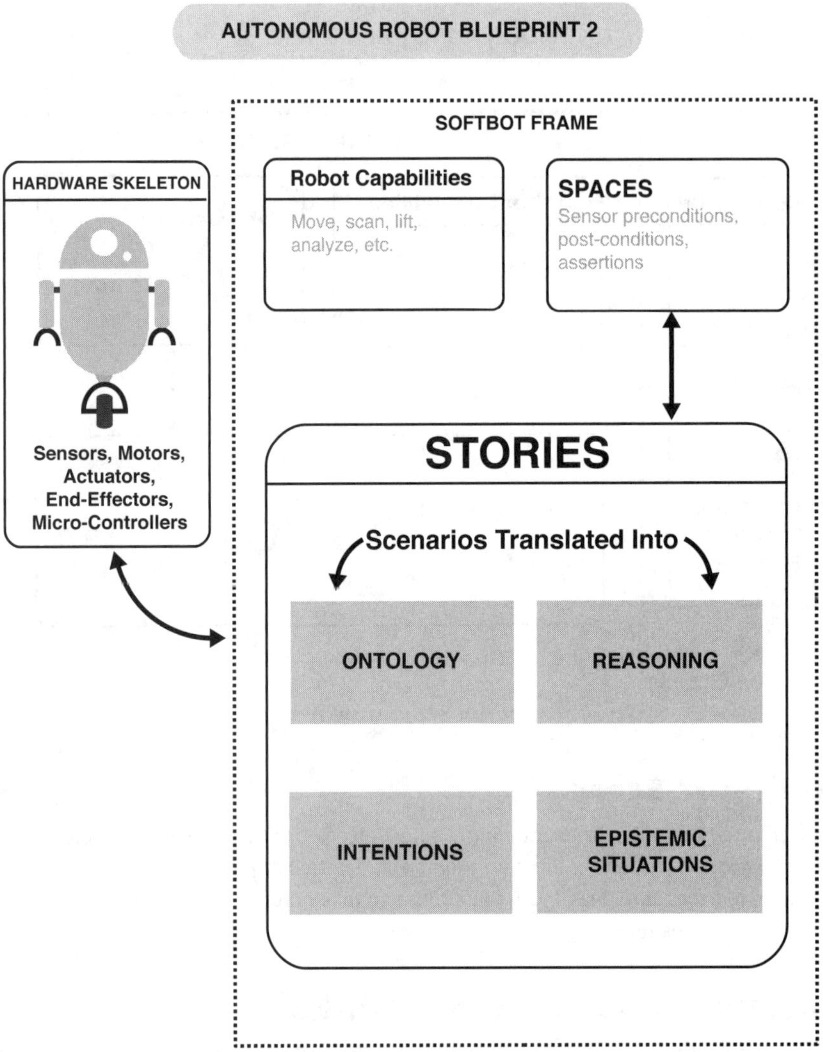

AUTONOMOUS ROBOT BLUEPRINT 2

Figure 10.2
Blueprint 2 of the autonomous robot anatomy: more detail

SOFTBOT FRAME

HARDWARE SKELETON

Robot Capabilities

Move, scan, lift, analyze, etc.

SPACES

Sensor preconditions, post-conditions, assertions

Sensors, Motors, Actuators, End-Effectors, Micro-Controllers

STORIES

Scenarios Translated Into

ONTOLOGY

REASONING

INTENTIONS

EPISTEMIC SITUATIONS

Table 10.3 A Simple Tchart Breakdown for Our Extended Robot Scenario

Things	Events	Actions
Robot	Leave original location	Locate object
Small room	Arrive at destination	Determine object color
Object	Return to original destination	Retrieve object

A Closer Look at the Scenario's Ontology

The things shown in Table 10.3 constitute a basic *ontology* of the extended robot scenario. For our purposes, an *ontology* is the set of things that make up a scenario, situation, or episode. It's important that we identify the ontology of the scenario for an autonomous robot. The ontology information for most teleoperated or remote-control robots is in the head and eyes of whoever is controlling the robot, which means that much of the detail of a situation can be left out of the robot's programming because the person with the controls is relying on her knowledge about the scenario.

For example, in some cases the robot doesn't have to worry about where it's going because the person with the controls is directing it. Details such as size, shape, or weight of an object can be left out because the person operating the robot can see how tall or how wide the object is and can direct the robot's end-effectors accordingly. In other cases, the robot's programming doesn't have to include what time to perform or not perform an action because the person doing the teleoperating pushes the start and stop button at the appropriate times. The more aspects of the robot that are under remote control means the robot needs less ontology specification. The more autonomy a robot has, the more ontology specification it will need.

In some approaches to robot autonomy, the programmer designs the robot's programming so that the robot can discover on its own the things in its scenario. This would be a level 4 or level 5 autonomous robot, as described in Table 8.1 in Chapter 8, "Getting Started with Autonomy: Building Your Robot's Softbot Counterpart." But for now, we focus on level 1 and level 2 autonomy and provide a simple and generic ontology of the scenario. But as you will see, where robots are concerned, the more detail you can provide about the things in the ontology, the better.

Providing Details for the "Things"

What size is the object in our scenario? Although we don't know the color (that's for the robot to determine), perhaps we know the object's size and weight, maybe even the object's shape. Where is the object located in the area? Exactly where is the robot's starting position in the area? Not only is it important to know which things make up a scenario, it is also important to know which of the details about those things will or should impact the robot's programming. Further, the details come in handy when we are trying to determine whether the robot can meet the REQUIRE specifications for the task.

It's useful to describe the things and the details about the things using terminology indicative of the scenario. If we know the object in Table 10.3 is a basketball, why not call it a basketball in the code? If the area is a gymnasium, we should use the term *gymnasium* instead of *area*. Breaking down the scenario into its list of things and naming them accordingly helps to come up with the robot's ROLL model used later for naming variables, procedures, functions, and methods. How much detail to provide is always a judgment call. Different programmers see the scenario from different points of view. But whichever approach is taken, keep in mind it is important to be consistent when describing details such as units of measurement, variables, and terminology. Don't switch back and forth. Figure 10.3 shows some of the details of the things in the extended robot's scenario.

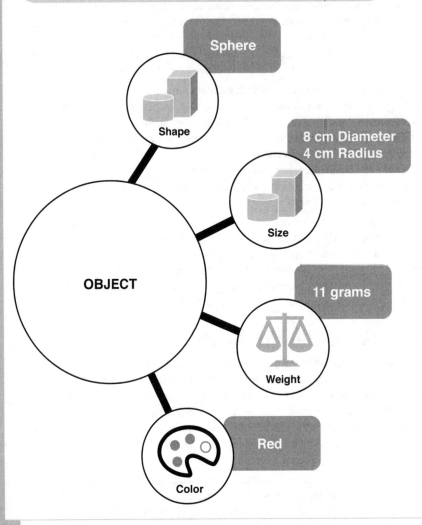

DETAILS OF THINGS IN THE EXTENDED ROBOT SCENARIO

Figure 10.3
Details of the things in the extended robot scenario

Using Object-Oriented Programming to Represent Things in the Scenario

In Chapter 8, we introduced the idea of the softbot frame and we demonstrated how classes and objects are used to represent the software component of a robot. All the things and actions in a robot scenario can also be represented with techniques of object-oriented programming using classes and methods. For example, in our extended robot scenario, we have five major classes listed in Table 10.4.

Table 10.4 Major Classes for Extended Robot Scenario

Class	Description
softbot **class**	Used to represent the robot
situation_object **class**	Used to represent objects in the situation
action methods	Used to represent actions the robot must perform in the situation
situation **class**	Used to represent various situations in the scenario
scenario **class**	Used to represent multiple scenarios and the robot's primary objectives

The details of the scenario are part of each class. Details of the situation are part of the situation_object, details of a softbot are part of the softbot object, and so on. The details can sometimes be represented by built-in data types such as strings, integers, or floating point numbers. Details can also be represented by classes. But these five classes represent the major components of our robot STORIES component. Figure 10.4 is a more detailed blueprint of our robot anatomy with the ontology component broken down into a little more detail.

Figure 10.4
More detailed blueprint of the robot anatomy with ontology component broken down

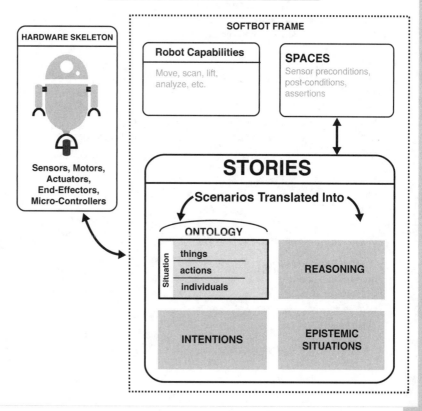

AUTONOMOUS ROBOT BLUEPRINT 3

SOFTBOT FRAME

HARDWARE SKELETON

Robot Capabilities
Move, scan, lift, analyze, etc.

SPACES
Sensor preconditions, post-conditions, assertions

Sensors, Motors, Actuators, End-Effectors, Micro-Controllers

STORIES

Scenarios Translated Into

ONTOLOGY

Situation

things
actions
individuals

REASONING

INTENTIONS

EPISTEMIC SITUATIONS

To illustrate how this works, we take a look at the Java code for our EV3 microcontroller and the C++ code for our Arduino microcontroller that make up the code for our Unit1 robot used in our extended robot scenario. BURT Translation Listing 10.1 shows the definition of the situation class.

BURT Translation Listing 10.1 Definition of the situation Class

BURT Translations Output: Java Implementation

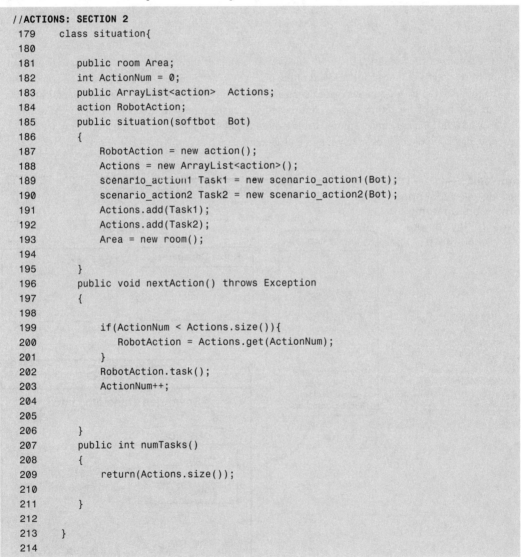

```
//ACTIONS: SECTION 2
179    class situation{
180
181        public room Area;
182        int ActionNum = 0;
183        public ArrayList<action>  Actions;
184        action RobotAction;
185        public situation(softbot  Bot)
186        {
187            RobotAction = new action();
188            Actions = new ArrayList<action>();
189            scenario_action1 Task1 = new scenario_action1(Bot);
190            scenario_action2 Task2 = new scenario_action2(Bot);
191            Actions.add(Task1);
192            Actions.add(Task2);
193            Area = new room();
194
195        }
196        public void nextAction() throws Exception
197        {
198
199            if(ActionNum < Actions.size()){
200                RobotAction = Actions.get(ActionNum);
201            }
202            RobotAction.task();
203            ActionNum++;
204
205
206        }
207        public int numTasks()
208        {
209            return(Actions.size());
210
211        }
212
213    }
214
```

The definition of the `situation` class starts on line 179. It contains a `room` object on line 181 and a list of action objects on line 183. It only has three methods:

- `constructor(situation())`

- `nextAction()`

- `numTasks()`

The `room` object is used to represent the location where the scenario situations take place. The actions list is used to store a list of action objects. There is one action object for each major task the robot has to perform. The constructor is defined on lines 185 to 195 and does most of the work for setting up the situation. The constructor creates two action objects (`Task1` and `Task2`) and creates a `room` object. Notice that the constructor adds the action objects to the list of actions. From this example we can see that so far there are two actions that the robot must execute in this situation. The `nextAction()` method causes the robot to execute the next action in the list.

```
if(ActionNum < Actions.size()){
    RobotAction = Actions.get(ActionNum);
    RobotAction.task();
    ActionNum++;
}
```

Notice the code on lines 199 to 203 that retrieves the next action in the `Actions` list and then uses

```
RobotAction.task();
```

as the method call to execute the `task()`. But how are `Task1` and `Task2` defined? Lines 189 to 190 refer to two scenario classes:

```
scenario_action1 Task1;
scenario_action2 Task2;
```

These classes inherit the `action` class that we created for this example. BURT Translation Listing 10.2 shows the definitions of the `action` class and the `scenario_action1` and `scenario_action2` classes.

BURT Translation Listing 10.2 Definitions of the `action`, `scenario_action1`, and `scenario_action2` Classes

BURT Translations Output: Java Implementation

```
//ACTIONS: SECTION 2
31      class action{
32          protected softbot Robot;
33          public action()
34          {   Robot = NULL;
35          }
36          public action(softbot Bot)
```

```
37      {
38          Robot = Bot;
39
40      }
41      public void task() throws Exception
42      {
43      }
44  }
45
46  class scenario_action1  extends action
47  {
48
49      public scenario_action1(softbot Bot)
50      {
51          super(Bot);
52      }
53      public void task() throws Exception
54      {
55          Robot.moveToObject();
56
57      }
58
59  }
60
61
62  class scenario_action2 extends action
63  {
64
65      public  scenario_action2(softbot Bot)
66      {
67
68          super(Bot);
69      }
70
71      public  void task() throws Exception
72      {
73          Robot.scanObject();
74
75      }
76
77
78  }
79
```

These classes are used to implement the notion of scenario actions. The action class is really used as a base class and we could have used a Java interface class here or a C++ abstract class (for Arduino implementation). But for now we want to keep things simple. So we use action as a base class, and `scenario_action1` and `scenario_action2` use `action` through inheritance. The most important method in the `action` classes is the `task()` method. This is the method that contains the code that the robot has to execute. Notice lines 53 to 57 and lines 71 to 75 make the calls to the code that represent the tasks:

```
53   public void task() throws Exception
54   {
55        Robot.moveToObject();
56
57   }
...
71   public  void task() throws Exception
72   {
73        Robot.scanObject();
74
75   }
```

These tasks cause the robot to move to the location of the object and determine the object's distance and color. Notice in Listing 10.1 and Listing 10.2 that there are no references to wires, pins, voltages, actuators, or effectors. This is an example of level 3 and above programming. At this level we try to represent the scenario naturally. Now of course we have to get to the actual motor and sensor code somewhere. What does `Robot.moveToObject()` actually do? How is `Robot.scanObject()` implemented? Both `moveToObject()` and `scanObject()` are methods that belong to the softbot frame that we named `softbot`. Let's look at `moveToObject()` first. BURT Translation Listing 10.3 shows the implementation code for `moveToObject()`.

BURT Translation Listing 10.3   Implementation Code for `moveToObject()`Method

BURT Translations Output: Java Implementation

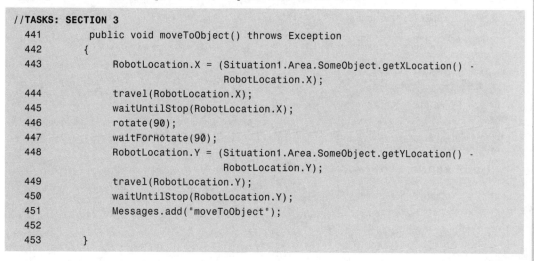

```
//TASKS: SECTION 3
441      public void moveToObject() throws Exception
442      {
443          RobotLocation.X = (Situation1.Area.SomeObject.getXLocation() -
                            RobotLocation.X);
444          travel(RobotLocation.X);
445          waitUntilStop(RobotLocation.X);
446          rotate(90);
447          waitForRotate(90);
448          RobotLocation.Y = (Situation1.Area.SomeObject.getYLocation() -
                            RobotLocation.Y);
449          travel(RobotLocation.Y);
450          waitUntilStop(RobotLocation.Y);
451          Messages.add("moveToObject");
452
453      }
```

This method gets the (X,Y) location coordinates that the robot needs to go to from the `Situation1` object on lines 443 and 448. This code illustrates that the robot's future X and Y locations are derived from the situation. According to lines 443 and 448 the situation has an area. The area has an object. And we get the object location by calling the methods:

```
SomeObject.getYLocation()
SomeObject.getXLocation()
```

Once we get those coordinates, we give the robot the command to `travel()` east or west the distance specified by X and north or south the distance specified by Y. Notice on line 181 from Listing 10.1 that the `situation` class has a `room` class. BURT Translation Listing 10.4 shows the definition of the `room` class.

BURT Translation Listing 10.4   Definition of the `room` Class

BURT Translations Output: Java Implementation

```
//Scenarios/Situations: SECTION 4
135    class room{
136        protected int Length;
137        protected int Width;
138        protected int Area;
139        public something SomeObject;
140
141        public  room()
142        {
143            Length = 300;
144            Width = 200;
145            SomeObject =  new something();
146            SomeObject.setLocation(20,50);
147        }
148        public int  area()
149        {
150            Area = Length * Width;
151            return(Area);
152        }
153
154        public  int length()
155        {
156
157            return(Length);
158        }
159
160        public int width()
161        {
162
163            return(Width);
164        }
165    }
```

We can see in Listing 10.4 that the room class has a something class. The room constructor sets the length of the room to 300 cm and the width of the room to 200 cm. It creates a new something object and sets its location (20,50) within the room. So the something class is used to represent the object and ultimately the object's location. Let's inspect the something class in BURT Translation Listing 10.5.

BURT Translation Listing 10.5 Implementation of the something Class

BURT Translations Output: Java Implementation

```
//Scenarios/Situations: SECTION 4
94   class something{
95       x_location Location;
96       int Color;
97       public something()
98       {
99           Location = new x_location();
100          Location.X = 0;
101          Location.Y = 0;
102          Color = 0;
103      }
104      public void setLocation(int X,int Y)
105      {
106
107          Location.X = X;
108          Location.Y = Y;
109
110      }
111      public int getXLocation()
112      {
113          return(Location.X);
114      }
115
116      public int getYLocation()
117      {
118          return(Location.Y);
119
120      }
121
122      public void setColor(int X)
123      {
124
125          Color = X;
126      }
127
128      public int getColor()
129      {
```

```
130            return(Color);
131        }
132
133    }
```

We see that something has a location and a color declared on lines 95 and 96. The pattern should be clear from Listing 10.1 through Listing 10.5; that is, we break down the robot's situation into a list of things and actions. We represent or "model" the things and actions using the notion of a class in some object-oriented language. The classes are then put together to form the robot's situation, and the situation is declared as a data member, property, or attribute of the robot's controller.

Decisions a Robot Can Make, Rules a Robot Can Follow

After you have identified the things and actions in the scenario (steps 1 and), it's time to identify the decisions and choices the robot will have concerning those things and actions. You must determine what course of action the robot will take and when based on the details of the scenario (step 3). If some of the robot's actions are optional, or if some of the robot's actions depend on what is found in the scenario, now is the time to identify decisions the robot will have to make. If there are certain courses of action that will always be made depending on certain conditions, now is the time to identify the rules that should apply when those conditions are met.

Once the details of the scenario are identified, it is appropriate to identify the pre/postconditions (SPACES) of the scenario. These decisions and rules make up the reasoning component of our robot STORIES structure. The reasoning component is an important part of robot autonomy. If the robot has no capacity to make decisions about things, or events, or actions that take place in the scenario, the robot's autonomy is severely limited. Robot decisions are implemented using the basic if-then-else, while-do, do-while, and case control structures. Recall from Listing 10.2 that the robot has a Task1 and a Task2. We have already seen the code moveToObject() that implements Task1. Task2 implements scanObject(), which is shown in BURT Translation Listing 10.6.

BURT Translation Listing 10.6   Implementation Code for scanObject

BURT Translation Output: Java Implementations

```
//TASKS: SECTION 3
455        public void scanObject()throws Exception
456        {
457
458            float Distance = 0;
459            resetArm();
460            moveSensorArray(110);
461            Thread.sleep(2000);
462            Distance = readUltrasonicSensor();
463            Thread.sleep(4000);
464            if(Distance <= 10.0){
```

```
465                getColor();
466                Thread.sleep(3000);
467            }
468        moveSensorArray(50);
469        Thread.sleep(2000);
470
471        }
```

There is a single simple decision for the robot to make on line 464. If the distance from the object is less than 10.0 centimeters, then the robot is instructed to determine the object's color. If the robot is more than 10 centimeters from the object, it will not attempt to determine the object's color. On line 462 the robot uses an ultrasonic sensor to measure the distance to the object. Once the robot has measured the distance it has a decision to make. Figure 10.5 is the robot anatomy blueprint that contains the reasoning component broken down into a bit more detail.

Figure 10.5
The robot anatomy blueprint that contains the reasoning component broken down.

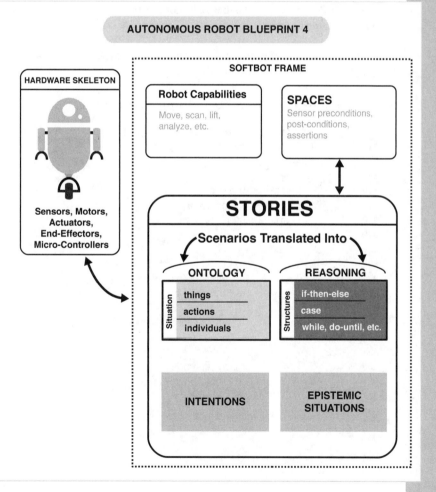

The idea is to set up the decisions so that each one either leads the robot to correctly execute the task or prevents the robot from taking unnecessary or impossible steps. Each decision should lead the robot closer to the completion of its primary task. Keep in mind that as the level of autonomy increases for the robot, the number and sometimes the complexity of the decision paths for the robot also increase.

Paying Attention to the Robot's Intention

Every situation has one or more actions. These actions together make up the robot's tasks. Each task represents one of the robot *intentions* (sometime referred to as *goals*). For instance, if the robot has four tasks, this means it has four intentions. Each decision that the robot makes should move it closer to one or more of its intentions. The program should give the robot explicit instructions if it cannot meet one or more of its intentions. If the robot cannot carry out its intentions, then it cannot fulfill its role in the scenario. Together the set of intentions represent the robot's role in the scenario. In C++ and Java the `main()` function is where the robot's primary intentions or primary tasks are placed. BURT Translation Listing 10.7 shows a snippet of the `main()` function for our extended robot scenario example.

BURT Translation Listing 10.7   The `main()` Function for Our Extended Robot Example

BURT Translation Output: Java Implementations

```
812        public static void main(String [] args)   throws Exception
813        {
814
815
816            softbot Unit1;
817            float Distance = 0;
818            int TaskNum = 0;
819
820            try{
821                Unit1 = new softbot();
822                TaskNum = Unit1.numTasks();
823                for(int N = 0; N < TaskNum; N++)
824                {
825                    Unit1.doNextTask();
826
827                }
828                Unit1.report();
829                Unit1.closeLog();
830
831            }
...

847
848        }
```

The `main()` function shows that our robot has a simple set of intentions. Line 822 shows that the robot is going to get the total number of tasks that it is supposed to execute. It then executes all the tasks in the situation. Lines 823 to 827 show that a simple loop structure controls how the robot approaches its intentions. After it is done executing the tasks it reports and then closes the log. In this oversimplified situation the robot simply attempts to execute the actions stored in the action list sequentially. However, for more sophisticated tasks, the way the next action is selected is typically based on robot decisions, the environment as detected by the sensors, the distance the robot has to travel, power considerations, time considerations, priority of tasks, SPACES that have or have not been met, and so on. On line 825 `Unit1` invokes the method `doNextTask()`. This method ultimately depends on the list of intentions (actions) that are part of the robot's situation. Figure 10.6 shows the detailed blueprint of our anatomy for autonomous robots and how the intentions component is just a breakdown of the robot's tasks.

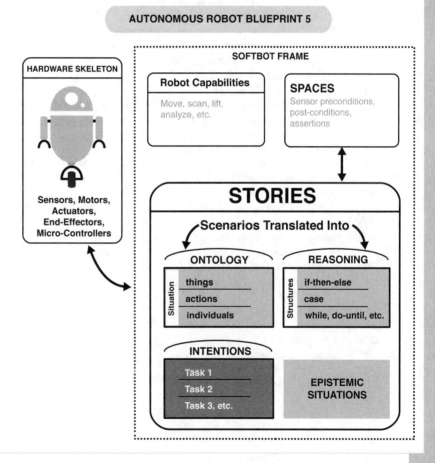

Figure 10.6
The robot anatomy blueprint that contains the intentions component broken down.

In Figure 10.6 the STORIES software component is clarified to show how the major components are represented or implemented. Once the robot's STORIES components are tied to its SPACES requirements and implemented, we have the basis for a robot that can carry out its tasks autonomously without the intervention of remote control. So far in our extended robot scenario we have used Java and the leJOS library to implement the STORIES components. BURT Translation Listing 10.8 contains most of the extended robot scenario program.

BURT Translation Listing 10.8 Implementation of the Extended Robot Scenario Program

BURT Translation Output: Java Implementations

```
3
4      import java.io.DataInputStream;
5      import java.io.DataOutputStream;
6      import lejos.hardware.sensor.NXTUltrasonicSensor;
7      import lejos.hardware.*;
8      import lejos.hardware.ev3.LocalEV3;
9      import lejos.hardware.port.SensorPort;
10     import lejos.hardware.sensor.SensorModes;
11     import lejos.hardware.port.Port;
12     import lejos.hardware.lcd.LCD;
13     import java.net.ServerSocket;
14     import java.net.Socket;
15     import lejos.hardware.sensor.HiTechnicColorSensor;
16     import lejos.hardware.sensor.EV3UltrasonicSensor;
17     import lejos.robotics.navigation.*;
18     import lejos.robotics.navigation.DifferentialPilot;
19     import lejos.robotics.localization.OdometryPoseProvider;
20     import lejos.robotics.SampleProvider;
21     import lejos.hardware.device.tetrix.*;
22     import lejos.hardware.device.tetrix.TetrixRegulatedMotor;
23     import lejos.robotics.navigation.Pose;
24     import lejos.robotics.navigation.Navigator;
25     import lejos.robotics.pathfinding.Path;
26     import java.lang.Math.*;
27     import java.io.PrintWriter;
28     import java.io.File;
29     import java.util.ArrayList;
30
//ACTIONS: SECTION 2
31     class action{
32        protected softbot Robot;
33        public action()
34        {
35        }
36        public action(softbot Bot)
37        {
38            Robot = Bot;
39
40        }
41        public void task() throws Exception
42        {
```

```
43        }
44    }
45
46    class scenario_action1  extends action
47    {
48
49        public scenario_action1(softbot Bot)
50        {
51            super(Bot);
52        }
53        public void task() throws Exception
54        {
55            Robot.moveToObject();
56
57        }
58
59    }
60
61
62    class scenario_action2 extends action
63    {
64
65        public  scenario_action2(softbot Bot)
66        {
67
68            super(Bot);
69        }
70
71        public  void task() throws Exception
72        {
73            Robot.scanObject();
74
75        }
76
77
78    }
79
//Scenario/Situation
80
81    class x_location{
82        public int X;
83        public int Y;
84        public x_location()
85        {
86
87            X = 0;
88            Y = 0;
```

```
 89        }
 90
 91    }
 92
 93
 94    class something{
 95       x_location Location;
 96       int Color;
 97       public something()
 98       {
 99            Location = new x_location();
100            Location.X = 0;
101            Location.Y = 0;
102            Color = 0;
103       }
104       public void setLocation(int X,int Y)
105       {
106
107            Location.X = X;
108            Location.Y = Y;
109
110       }
111       public int getXLocation()
112       {
113            return(Location.X);
114       }
115
116       public int getYLocation()
117       {
118            return(Location.Y);
119
120       }
121
122       public void setColor(int X)
123       {
124
125            Color = X;
126       }
127
128       public int getColor()
129       {
130            return(Color);
131       }
132
133    }
134
135    class room{
```

```
136       protected int Length = 300;
137       protected int Width = 200;
138       protected int Area;
139       public something SomeObject;
140
141       public  room()
142       {
143           SomeObject =  new something();
144           SomeObject.setLocation(20,50);
145
146
147       }
148       public int  area()
149       {
150           Area = Length * Width;
151           return(Area);
152       }
153
154       public  int length()
155       {
156
157           return(Length);
158       }
159
160       public int width()
161       {
162
163           return(Width);
164       }
165   }
166
167   class situation{
168
169       public room Area;
170       public situation()
171       {
172           Area = new room();
173
174       }
175
176   }
177
178
179   class situation{
180
181       public room Area;
182       int ActionNum = 0;
```

```
183        public ArrayList<action>  Actions;
184        action RobotAction;
185        public situation(softbot  Bot)
186        {
187            RobotAction = new action();
188            Actions = new ArrayList<action>();
189            scenario_action1 Task1 = new scenario_action1(Bot);
190            scenario_action2 Task2 = new scenario_action2(Bot);
191            Actions.add(Task1);
192            Actions.add(Task2);
193            Area = new room();
194
195        }
196        public void nextAction() throws Exception
197        {
198
199            if(ActionNum < Actions.size()){
200                RobotAction = Actions.get(ActionNum);
201            }
202            RobotAction.task();
203            ActionNum++;
204
205
206        }
207        public int numTasks()
208        {
209            return(Actions.size());
210
211        }
212
213    }
214
215    public class softbot
216    {
//PARTS: SECTION 1
//Sensor Section
217        public EV3UltrasonicSensor Vision;
218        public HiTechnicColorSensor ColorVision;

219        int CurrentColor;
220        double  WheelDiameter;
221        double TrackWidth;
222        float  RobotLength;
223        DifferentialPilot  D1R1Pilot;
224        ArcMoveController  D1R1ArcPilot;
//Actuators
225        TetrixControllerFactory  CF;
```

```
226        TetrixMotorController MC;
227        TetrixServoController SC;
228        TetrixRegulatedMotor LeftMotor;
229        TetrixRegulatedMotor RightMotor;
230        TetrixServo  Arm;
231        TetrixServo  Gripper;
232        TetrixServo  SensorArray;
//Support
233        OdometryPoseProvider Odometer;
234        Navigator D1R1Navigator;
235        boolean PathReady = false;
236        Pose CurrPos;
237        int OneSecond = 1000;
238        Sound  AudibleStatus;
239        DataInputStream dis;
240        DataOutputStream Dout;
241        location  CurrentLocation;
242        SampleProvider UltrasonicSample;
243        SensorModes USensor;
244        PrintWriter Log;


//Situations/Scenarios: SECTION 4
245        situation Situation1;
246        x_location RobotLocation;


247        ArrayList<String> Messages;
248        Exception SoftbotError;
249
250        public softbot() throws InterruptedException,Exception
251        {
252
253            Messages = new ArrayList<String>();
254            Vision = new EV3UltrasonicSensor(SensorPort.S3);
255            if(Vision == null){
256                Messages.add("Could Not Initialize Ultrasonic Sensor");
257                SoftbotError = new Exception("101");
258                throw SoftbotError;
259            }
260            Vision.enable();
261            Situation1 = new situation(this);
262            RobotLocation = new x_location();
263            RobotLocation.X = 0;
264            RobotLocation.Y = 0;
265
266            ColorVision = new HiTechnicColorSensor(SensorPort.S2);
267            if(ColorVision == null){
268                Messages.add("Could Not Initialize Color Sensor");
```

```
269                 SoftbotError = new Exception("100");
270                 throw SoftbotError;
271             }
272         Log = new PrintWriter("softbot.log");
273         Log.println("Sensors  constructed");
274         Thread.sleep(1000);
275         WheelDiameter = 7.50f;
276         TrackWidth = 32.5f;
277
278         Port APort = LocalEV3.get().getPort("S1");
279         CF = new TetrixControllerFactory(SensorPort.S1);
280         if(CF == null){
281             Messages.add("Could Not Setup Servo Port");
282             SoftbotError = new Exception("102");
283             throw SoftbotError;
284         }
285         Log.println("Tetrix Controller Factor Constructed");
286
287         MC = CF.newMotorController();
288         SC = CF.newServoController();
289
290         LeftMotor = MC.getRegulatedMotor(TetrixMotorController.MOTOR_1);
291         RightMotor = MC.getRegulatedMotor(TetrixMotorController.MOTOR_2);
292         if(LeftMotor == null ¦¦ RightMotor == null){
293             Messages.add("Could Not Initalize Motors");
294             SoftbotError = new Exception("103");
295             throw SoftbotError;
296         }
297         LeftMotor.setReverse(true);
298         RightMotor.setReverse(false);
299         LeftMotor.resetTachoCount();
300         RightMotor.resetTachoCount();
301         Log.println("motors Constructed");
302         Thread.sleep(2000);
303
304
317         SensorArray = SC.getServo(TetrixServoController.SERVO_3);
318         if(SensorArray == null){
319             Messages.add("Could Not Initialize SensorArray");
320             SoftbotError = new Exception("107");
321             throw SoftbotError;
322         }
323         Messages.add("Servos Constructed");
324         Log.println("Servos Constructed");
325         Thread.sleep(1000);
326
327         SC.setStepTime(7);
```

```
328             Arm.setRange(750,2250,180);
329             Arm.setAngle(100);
330             Thread.sleep(1000);
331
335
336
337             SensorArray.setRange(750,2250,180);
338             SensorArray.setAngle(20);
339             Thread.sleep(1000);
340             D1R1Pilot = new DifferentialPilot
                              (WheelDiameter,TrackWidth,LeftMotor,RightMotor);
341             D1R1Pilot.reset();
342             D1R1Pilot.setTravelSpeed(10);
343             D1R1Pilot.setRotateSpeed(20);
344             D1R1Pilot.setMinRadius(0);
345
346             Log.println("Pilot Constructed");
347             Thread.sleep(1000);
348             CurrPos = new Pose();
349             CurrPos.setLocation(0,0);
350             Odometer = new OdometryPoseProvider(D1R1Pilot);
351             Odometer.setPose(CurrPos);
352
353
354             D1R1Navigator = new Navigator(D1R1Pilot);
355             D1R1Navigator.singleStep(true);
356             Log.println("Odometer Constructed");
357
358             Log.println("Room  Width: " + Situation1.Area.width());
359             room SomeRoom = Situation1.Area;
360             Log.println("Room Location: " +
                              SomeRoom.SomeObject.getXLocation() + "," +
                              SomeRoom.SomeObject.getYLocation());
361             Messages.add("Softbot Constructed");
362             Thread.sleep(1000);
363
364
365
366         }
367
434
//TASKS: SECTION 3
435         public void doNextTask() throws Exception
436         {
437             Situation1.nextAction();
438
439         }
```

```
440
441     public void moveToObject() throws Exception
442     {
443         RobotLocation.X = (Situation1.Area.SomeObject.getXLocation()
                                - RobotLocation.X);
444         travel(RobotLocation.X);
445         waitUntilStop(RobotLocation.X);
446         rotate(90);
447         waitForRotate(90);
448         RobotLocation.Y = (Situation1.Area.SomeObject.getYLocation()
                                - RobotLocation.Y);
449         travel(RobotLocation.Y);
450         waitUntilStop(RobotLocation.Y);
451         Messages.add("moveToObject");
452
453     }
454
455     public void scanObject()throws Exception
456     {
457
458         float Distance = 0;
459         resetArm();
460         moveSensorArray(110);
461         Thread.sleep(2000);
462         Distance = readUltrasonicSensor();
463         Thread.sleep(4000);
464         if(Distance <= 10.0){
465             getColor();
466             Thread.sleep(3000);
467         }
468         moveSensorArray(50);
469         Thread.sleep(2000);
470
471     }
472
473     public int numTasks()
474     {
475         return(Situation1.numTasks());
476     }
477
512
513     public float readUltrasonicSensor()
514     {
515         UltrasonicSample =  Vision.getDistanceMode();
516         float X[] = new float[UltrasonicSample.sampleSize()];
517         Log.print("sample size ");
518         Log.println(UltrasonicSample.sampleSize());
```

```
519
520         UltrasonicSample.fetchSample(X,0);
521         int Line = 3;
522         for(int N = 0; N < UltrasonicSample.sampleSize();N++)
523         {
524
525             Float Temp = new Float(X[N]);
526             Log.println(Temp.intValue());
527             Messages.add(Temp.toString());
528             Line++;
529
530         }
531         if(UltrasonicSample.sampleSize() >= 1){
532             return(X[0]);
533         }
534         else{
535                 return(-1.0f);
536         }
537
538     }
539     public int getColor()
540     {
541
542         return(ColorVision.getColorID());
543     }
544
545
546     public void identifyColor() throws Exception
547     {
548         LCD.clear();
549         LCD.drawString("color identified",0,3);
550         LCD.drawInt(getColor(),0,4);
551         Log.println("Color Identified");
552         Log.println("color = " + getColor());
553     }
554     public void rotate(int Degrees)
555     {
556         D1R1Pilot.rotate(Degrees);
667     }
558     public void forward()
559     {
560
561         D1R1Pilot.forward();
562     }
563     public void backward()
564     {
565         D1R1Pilot.backward();
```

```
566        }
567
568        public void travel(int Centimeters)
569        {
570            D1R1Pilot.travel(Centimeters);
571
572        }
573
601        public void moveSensorArray(float X) throws Exception
602        {
603
604            SensorArray.setAngle(X);
605            while(SC.isMoving())
606            {
607                Thread.sleep(1500);
608            }
609
610
611        }
612
641
642        public boolean waitForStop()
643        {
644            return(D1R1Navigator.waitForStop());
645
646        }
647
648
703
704
705
706        public void waitUntilStop(int Distance) throws Exception
707        {
708
709            Distance = Math.abs(Distance);
710            Double  TravelUnit = new Double
                                (Distance/D1R1Pilot.getTravelSpeed());
711            Thread.sleep(Math.round(TravelUnit.doubleValue()) * OneSecond);
712            D1R1Pilot.stop();
713            Log.println("Travel Speed " + D1R1Pilot.getTravelSpeed());
714            Log.println("Distance:   " + Distance);
715            Log.println("Travel Unit: " + TravelUnit);
716            Log.println("Wait for: " + Math.round
                        (TravelUnit.doubleValue()) * OneSecond);
717
718        }
719        public void waitUntilStop()
```

```
720            {
721                do{
722
723                }while(D1R1Pilot.isMoving());
724                D1R1Pilot.stop();
725
726            }
727
728            public void waitForRotate(double Degrees) throws Exception
729            {
730
731                Degrees = Math.abs(Degrees);
732                Double DegreeUnit = new Double
                                   (Degrees/D1R1Pilot.getRotateSpeed());
733                Thread.sleep(Math.round(DegreeUnit.doubleValue()) * OneSecond);
734                D1R1Pilot.stop();
735                Log.println("Rotate Unit: " + DegreeUnit);
736                Log.println("Wait for: " + Math.round
                            (DegreeUnit.doubleValue()) * OneSecond);
737
738
739
740            }
741
742
743
744            public  void closeLog()
745            {
746                Log.close();
747            }
748
749
750            public void report() throws Exception
751            {
752                ServerSocket Client = new ServerSocket(1111);
753                Socket SomeSocket = Client.accept();
754                DataOutputStream Dout = new DataOutputStream
                                   (SomeSocket.getOutputStream());
755                Dout.writeInt(Messages.size());
756                for(int N = 0;N < Messages.size();N++)
757                {
758                    Dout.writeUTF(Messages.get(N));
759                    Dout.flush();
760                    Thread.sleep(1000);
761
762                }
763                Thread.sleep(1000);
```

```
764                 Dout.close();
765                 Client.close();
766             }
767
768
769         public void report(int X) throws Exception
770             {
771
772                 ServerSocket Client = new ServerSocket(1111);
773                 Socket SomeSocket = Client.accept();
774                 DataOutputStream Dout = new DataOutputStream
                                        (SomeSocket.getOutputStream());
775                 Dout.writeInt(X);
776                 Thread.sleep(5000);
777                 Dout.close();
778                 Client.close();
779
780
781             }
782         public void addMessage(String X)
783             {
784                 Messages.add(X);
785             }
786
812         public static void main(String [] args)   throws Exception
813             {
814
815
816                 softbot Unit1;
817                 float Distance = 0;
818                 int TaskNum = 0;
819
820                 try{
821                     Unit1 = new softbot();
822                     TaskNum = Unit1.numTasks();
823                     for(int N = 0; N < TaskNum; N++)
824                     {
825                         Unit1.doNextTask();
826
827                     }
828                     Unit1.report();
829                     Unit1.closeLog();
830
831                 }
832                 catch(Exception E)
833                 {
834                     Integer Error;
```

```
835                         System.out.println("Error is : " + E);
836                         Error = new Integer(0);
837                         int RetCode = Error.intValue();
838                         if(RetCode == 0){
839                             RetCode = 999;
840                         }
841                         ServerSocket Client = new ServerSocket(1111);
842                         Socket SomeSocket = Client.accept();
843                         DataOutputStream Dout = new DataOutputStream
                                            (SomeSocket.getOutputStream());
844                         Dout.writeInt(RetCode);
845                         Dout.close();
846                         Client.close();
847
848
849                 }
850
851
852
853         }
854
855
856
857
858     }
```

But recall in our extended scenario, one of the tasks the robot has to execute is to retrieve the object. Our robot build uses the PhantomX Pincher Arm from Trossen Robotics, and it is based on the Arbotix controller, which is an Arduino compatible platform. Our robot has more than one microcontroller. We use serial and Bluetooth connections to communicate between the microcontrollers. We show some of the communication details in Chapter 11, "Putting It All Together: How Midamba Programmed His First Autonomous Robot." Figure 10.7 shows a photo of an Arduino/EV3-based robot.

There are two arms on the robot, each having a different degree of freedom and gripper types as discussed in Chapter 7. The PhantomX Pincher and the Tetrix-based robot arms are highlighted in Figure 10.7. The BURT Translation input shown in Listing 10.9 shows some of the basic activities the Arduino-based Arbotix controller has to perform.

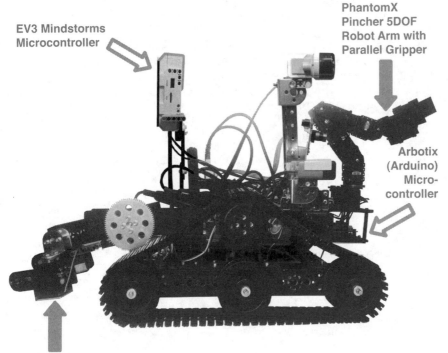

UNIT1 (ARDUINO & EV3-BASED) ROBOT

Figure 10.7
A photo of an Arduino EV3-based robot

EV3 Mindstorms
Microcontroller

PhantomX
Pincher 5DOF
Robot Arm with
Parallel Gripper

Arbotix
(Arduino)
Micro-
controller

Terix-based 1DOF
Robot Arm with
Angular Gripper

BURT Translation Listing 10.9   Some Basic Activities the Arduino-Based Arbotix Controller Performs

BURT Translation INPUT

```
Softbot   Frame
Name:  Unit1
Parts:
Actuator Section:
Servo and its gripper (for movement)


Actions:
Step 1: Check the voltage going to the servo
Step 2: Check each servo to see if it's operating and has the correct starting position
Step 3: Set the servos to a center position
```

Step 4: Set the position of a servo

Step 5: Open the gripper

Step 6: Close the gripper

Tasks:

Position the servo of the gripper in order to open and close the gripper.

End Frame

These are some basic activities that any robot arm controller component would perform. One of the robots that we use for our examples has arms with different controllers and different software but both arms perform basically the same activities. For the EV3, we build our robot arm code on top of the leJOS Java class libraries. For PhantomX, we build our code on top of Bioloid class libraries and AX Dynamixel code. Recall that our STORIES structure includes the things and actions that are part of the scenario, and we use the notion of object-oriented classes to represent the things and the actions. Keep in mind that the robot is one of the things in the scenario, so we also represent the robot and its components using object-oriented classes. The complete BURT Translation for the robot arm capability is shown in Listing 10.10.

BURT Translation Listing 10.10   Some Robot Arm Capabilities

BURT Translations Output: C++ Implementations

```
1     #include <ax12.h>
2     #include <BioloidController.h>
3     #include "poses.h"
4
5     BioloidController bioloid = BioloidController(1000000);
6
7     class robot_arm{
8        private:
9           int ServoCount;
10       protected:
11          int id;
12          int pos;
13          boolean IDCheck;
14          boolean StartupComplete;
15       public:
16          robot_arm(void);
...
//ACTIONS: SECTION 2
18             void scanServo(void);
19             void moveCenter(void);
20             void checkVoltage(void);
```

```
21          void moveHome(void);
22          void relaxServos(void);
23          void retrieveObject(void);
24      };
25

...
//TASKS: SECTION 3
82      void robot_arm::scanServo(void)
83      {
84          id = 1;
85          Serial.println("Scanning Servo.....");
86          while (id <= ServoCount)
87          {
88              pos =  ax12GetRegister(id, 36, 2);
89              Serial.print("Servo ID: ");
90              Serial.println(id);
91              Serial.print("Servo Position: ");
92              Serial.println(pos);
93              if (pos <= 0){
94                  Serial.println("==============================");
95                  Serial.print("ERROR! Servo ID: ");
96                  Serial.print(id);
97                  Serial.println(" not found. Please check connection and
                                        verify correct ID is set.");
98                  Serial.println("==============================");
99                  IDCheck = false;
100             }
101
102             id = (id++)%ServoCount;
103             delay(1000);
104         }
105         if (!IDCheck){
106             Serial.println("=================================");
107             Serial.println("ERROR! Servo ID(s) are missing from Scan.");
108             Serial.println("=================================");
109         }
110         else{
111                 Serial.println("Servo Check Passed");
112         }
...

223     void robot_arm::checkVoltage(void)
224     {
225         float voltage = (ax12GetRegister (1, AX_PRESENT_VOLTAGE, 1)) / 10.0;
226         Serial.print ("System Voltage: ");
227         Serial.print (voltage);
228         Serial.println (" volts.");
```

```
229        if (voltage < 10.0){
230            Serial.println("Voltage levels below 10v, please charge battery.");
231            while(1);
232        }
233        if (voltage > 10.0){
234            Serial.println("Voltage levels nominal.");
235        }
236        if (StartupComplete){
237                ...
238        }
239
240    }
241
242    void robot_arm::moveCenter()
243    {
244        delay(100);
245        bioloid.loadPose(Center);
246        bioloid.readPose();
247        Serial.println("Moving servos to centered position");
248        delay(1000);
249        bioloid.interpolateSetup(1000);
250        while(bioloid.interpolating > 0){
251                bioloid.interpolateStep();
252                delay(3);
253        }
254        if (StartupComplete){
255            …
256        }
257    }
258
259
260    void robot_arm::moveHome(void)
261    {
262        delay(100);
263        bioloid.loadPose(Home);
264        bioloid.readPose();
265        Serial.println("Moving servos to Home position");
266        delay(1000);
267        bioloid.interpolateSetup(1000);
268        while(bioloid.interpolating > 0){
269                bioloid.interpolateStep();
270                delay(3);
271        }
272        if (StartupComplete){
273            …
274        }
275    }
```

```
...
277    void robot_arm::retrieveObject()
278    {
279
280        Serial.println("=======================");
281        Serial.println("Retrieve Object");
282        Serial.println("=======================");
283        delay(500);
284        id  = 1;
285        pos = 512;
286
287
288        Serial.print(" Adjusting Servo : ");
289        Serial.println(id);
290        while(pos >= 312)
291        {
292                SetPosition(id,pos);
293                pos = pos--;
294                delay(10);
295        }
296        while(pos <= 512){
297                SetPosition(id, pos);
298                pos = pos++;
299                delay(10);
300        }
301
302
303        id = 3;
304        Serial.print("Adjusting Servo ");
305        Serial.println(id);
306        while(pos >= 200)
307        {
308                SetPosition(id,pos);
309                pos = pos--;
310                delay(15);
311        }
312        while(pos <= 512){
313                SetPosition(id, pos);
314                pos = pos++;
315                delay(15);
316        }
317
318        id = 3;
319        while(pos >= 175)
320        {
321                SetPosition(id,pos);
322                pos = pos--;
```

```
323          delay(20);
324      }
325
326
327      id = 5;
328      Serial.print(" Adjusting Gripper : ");
329      Serial.println(id);
330      pos = 512;
331      while(pos >= 170)
332      {
333              SetPosition(id,pos);
334              pos = pos--;
335              delay(30);
336      }
337      while(pos <= 512){
338              SetPosition(id, pos);
339              pos = pos++;
340              delay(30);
341      }
342      // id 5 is the gripper
343      id = 5;
344      while(pos >= 175)
345      {
346              SetPosition(id,pos);
347              pos = pos--;
348              delay(20);
349      }
350      while(pos <= 512){
351              SetPosition(id, pos);
352              pos = pos++;
353              delay(20);
354      }
355      id = 4;
356      Serial.print(" Adjusting Servo : ");
357      Serial.println(id);
358
359      while(pos >= 200)
360      {
361              SetPosition(id,pos);
362              pos = pos--;
363              delay(10);
364      }
365      while(pos <= 512){
366              SetPosition(id, pos);
367              pos = pos++;
368              delay(20);
369      }
```

```
370
371          if(StartupComplete == 1){
372              ...
373          }
374
375      }
376
377
```

Listing 10.10 contains a partial C++ class declaration for our PhantomX Pincher robot arm and implementations for some of the major methods. Listing 10.10 shows how the method implementations are built on top of the Bioloid class methods. Note that we are using the Arduino Serial object to communicate what the arm is doing at any point.

For example, line 85 simply shows that we are getting ready to start the servo scanning process. We show the complete code for the robot arm in Chapter 11.

The constructor, which is not shown in Listing 10.10, sets up a baud rate of 9600 and a connection between the arm and the serial port on the microcontroller. We use the Bioloid class by making method calls. For example:

```
245          bioloid.loadPose(Center);
246          bioloid.readPose();
```

The commands on lines 245 and 246 are used to center the servos. The `retrieveObject()` method is implemented on lines 277 to 377 and shows examples of setting the positions of the AX servos (this robot arm has five servos), including opening and closing the gripper (Servo 5) using the `SetPosition()` as shown on lines 344 to 349:

```
344          while(pos >= 175)
345          {
346              SetPosition(id,pos);
347              pos = pos--;
348              delay(20);
349          }
```

> **note**
>
> All the components—sensors, motors, actuators, servos, and end-effectors—are implemented as object-oriented classes.

Object-Oriented Robot Code and Efficiency Concerns

You may be wondering whether there is an extra cost for using an object-oriented approach to programming a robot to perform autonomously versus not using an object-oriented approach. This has been a long and hard battle that is not likely to be won anytime soon. The proponents of the C language, or microcontroller assembly, are quick to point out how small the code footprint is when using their languages or how fast it is compared to something like Java, Python, or C++. They may have a point if you aren't concerned about modeling the environment the robot has to work in or the scenario and situation the robot has to perform in.

Designing and building maintainable, extensible, and understandable environmental and scenario models in microcontroller assembler or in C is considerably more difficult than using the object-oriented approach. So it's a matter of how you measure cost. If the size of the robot program or the absolute speed is the only concern, then a well optimized C program or microcontroller assembly program is hard to beat (although it could be matched in C++).

However, if the goal is to represent in software the robot's environment, scenario, situations, intentions, and reasoning (which is done in most autonomous approaches), then the object-oriented approach is hard to beat. A well-designed robot class, situation, and environment classes coded in a way that is easier to understand, change, and extend is rewarding. Also, the bark is sometimes worse than the bite. For instance, Table 10.5 shows the Java classes and byte-code sizes for each class that is necessary for our extended robot scenario.

Table 10.5   The Java Classes and Byte-Code Sizes for the Extended Robot Scenario

Class	Byte-Code Size (in kilobytes)
softbot	14081
room	596
scenario_action1	380
situation	1020
something	698
x_location	261
scenario_action2	376

The combined size of the classes that need to be uploaded to the EV3 microcontroller is 17412 kb, a little less than 18 k. The EV3 microcontroller has 64 MB of main RAM and 16 MB of flash. So our object-oriented approach to the extended robot scenario with all the STORIES and SPACES components barely scratches the surface. However, it would be accurate to say that our simple extended robot scenario is typical of a full-blown robot implementation of an autonomous task. Size and speed are definitely real concerns of the code. But the trade-off of space and speed for the power of expression in the object-oriented approach is well justified.

 tip

After step 1 is completed from Table 10.2, it's a good time to verify whether the robot actually has the capability to interact with the things and perform the actions specified in the ontology. In some cases, it is clear whether the robot is up to the task before step 1 is completed. If you're not certain whether the robot is up to the task, checking the robot's functionality after breaking down the list of things in the scenario serves as a good checkpoint. Otherwise, effort could be wasted programming a robot to do something that it simply cannot do.

What's Ahead?

In Chapter 11 we will discuss how the techniques presented in this book are used to solve Midamba's Predicament.

PUTTING IT ALL TOGETHER: HOW MIDAMBA PROGRAMMED HIS FIRST AUTONOMOUS ROBOT

Robot Sensitivity Training Lesson #11: *A robot needs power to maintain control.*

We started our robotic journey with poor, unfortunate Midamba who found himself marooned on a deserted island. Let's review Midamba's predicament.

Midamba's Initial Scenario

When we last saw Midamba in our robot boot camp his electric-powered Sea-Doo had run low on battery power. Midamba had a spare battery, but the spare had sat for a long time and had started to leak. The battery had corrosion on its terminals and was not working properly.

With one dying battery and one corroded battery, Midamba had just enough power to get to a nearby island where he might possibly find help. Unfortunately for Midamba, the only thing on the island was an experimental chemical facility totally controlled by autonomous robots.

Midamba figured if there was some kind of chemical in the facility that could neutralize the corrosion, he could clean his spare battery and be on his way. There was a front office to the facility, but it was separated from the warehouse where most of the chemicals were stored, and the only entrance to the warehouse was locked.

Midamba could see robots on the monitors moving about in the warehouse transporting containers, marking containers, lifting objects, and so on, but there was no way to get in. All that was in the front office was a computer, a microphone, and a manual titled *How to Program Autonomous Robots* by Cameron Hughes and Tracey Hughes. There were also a few robots and some containers and beakers, but they were all props. Perhaps with a little luck Midamba could find something in the manual that would allow him to instruct one of the robots to find and retrieve the chemical he needed.

Midamba Becomes a Robot Programmer Overnight!

Midamba finished reading the book and understood all he could. Now all that was left was to somehow get the robots to do his bidding. He watched the robots work and then stop at what appeared to be some kind of program reload station. He noticed each robot had a bright yellow and black label: Unit1, Unit2, and so on. Also, each robot seemed to have different capabilities. So Midamba quickly sketched out his situation.

Midamba's Situation #1

"I'm in what appears to be a control room. The robots are in a warehouse full of chemicals, and some of those chemicals can probably help me neutralize the corrosion on my battery. But I'm not quite sure what kind of corrosion I have. If the battery is alkaline-based, I will need some kind of acidic chemical to clean the corrosion. If the battery is nickel-zinc or nickel-cadmium-based, I will need some kind of alkaline or base chemical to clean the corrosion. If I'm lucky there might even be some kind of battery charger in the warehouse. According to the book, each of these robots must have a language and a specific set of capabilities. If I could find out what each robot's capabilities are and what language each robot uses, maybe I could reprogram some of them to look for the material I need. So first I have to find out what languages the robots use and second find out what capabilities the robots have."

As Midamba viewed each robot he recognized grippers and end-effectors that he saw in the book. He noticed that some robots were using what must be distance measuring sensors, and some robots appeared to be sampling or testing chemicals in the warehouse. Other robots seemed to be sorting containers based on colors or special marking. But that wasn't enough information. So Midamba ransacked the control room looking for more information about the robots, and as luck would have it, he hit the jackpot! He found the capability matrix for each robot. After a quick scan he was especially interested in the capabilities of the robots labeled Unit1 and Unit2. Table 11.1 is the capability matrix for Unit1 and Unit2.

Table 11.1 Capability Matrix for Unit1 and Unit2

Robot Name	Microcontroller/ Controllers/Processors/OS	Sensors/Actuators	End effectors	Mobility	Communication
Unit1 Tetrix-Based Robot	ARM9 (Java) ■ Linux OS ■ 300 MHz ■ 16 MB Flash ■ 64 MB RAM Arbotix (Arduino Uno) Spark Fun Red Board (Arduino Uno) 1 HiTechnic Servo Controller 1 HiTechnic DC Controller	Sensor Array ■ 1 Color Light ■ 1 Ultrasonic Touch (Gripper) Smartphone Camera 2 DC Motors 1 servo (sensor array)	Front Right Arm - 6 DOF PhantomX Pincher w/Linear Gripper Back Left Arm - 1 DOF w/ Angular Gripper	Tractor Wheeled	USB Port Bluetooth
Unit2 RS Media	200 MHz ARM9 with 64 MB of flash RAM Linux OS 16-bit processor 32 MB RAM	3 infrared (IR) detectors VGA quality camera 3 sound sensors for sound 2 touch sensors on the back of each hand 1 toe and 1 heel touch sensor in each foot 12 motors 12 DOF LCD screen Speakers Microphone	2 arms with 3 digit hand end-effectors	Bipedal	USB

The robots had ARM7, ARM9, and Arduino UNO controllers. Midamba could see that Unit2 had distance and color sensors. Unit1 had robotic arms, cameras, and some kind of chemical measurement capabilities.

He used the matrix to see what languages the robots could be programmed in and discovered that most of the robots could use languages such as Java and C++. Now that Midamba knew what languages were involved and what the robots' basic capabilities were, all he needed to do was put together some programming that would allow the robots to get those batteries working!

Midamba quickly read the book and managed to get some of the basics for programming robots to execute a task autonomously. As far as he could see, the basic process boiled down to five steps:

1. Write out the complete scenario that includes the robot's role in simple-to-understand language, making sure that the robots have the capabilities to execute that role.

2. Develop a ROLL model to use for programming the robot.

3. Make an RSVP for the scenario.

4. Identify the robot's SPACES from the RSVP.

5. Develop and upload the appropriate STORIES to the robot.

Although he didn't understand everything in the book, these did seem to be the basic steps. So we will follow Midamba and these five steps and see where it gets us.

Step 1. Robots in the Warehouse Scenario

In a nutshell, Midamba's solution to the problem requires a robot to give him a basic inventory of what kind of chemicals are in the warehouse. After he gets the inventory, he can determine which (if any) chemicals could be useful in neutralizing the battery corrosion. If there are potentially useful chemicals, he needs to verify that by having a robot analyze the chemical and then in some way have the robot deliver the chemical to him so it can be used. The basic scenario can be summarized using the following simple language.

Midamba's Facility Scenario #1

Using one or more robots in the facility, scan the facility and return an inventory of the containers that are found. Determine whether any of the containers hold substances that can be used to neutralize and clean the kind of corrosion that can occur with alkaline or nickel-based batteries. If such containers are found, retrieve them and return them to the front office.

This type of high-level summary is the first step in the process. It clarifies what the main objectives are and what role the robot(s) will play in accomplishing those objectives. Once you have this type of high-level summary of the scenario, it is useful to either consult an existing robot capability matrix or construct one to immediately see whether the robots meet the REQUIRE aspects of the scenario. If they don't, they must be improved or you must stop at this point realizing that the robot(s) will not be able to complete the tasks. The whole point of writing this type of description is so that it is clear what the goals are and exactly what to expect the robot to do. Ultimately, giving the robot simple instructions that would accomplish the task is the goal in this situation. For example:

Robot, go find something that will clean the corrosion off the battery.

In fact, that's precisely the command Midamba wants to give the robot. But robots don't yet understand language at this level. So the ROLL model is used as a translation mechanism to translate between the human language used to express the initial instructions and the language the robot will ultimately use to execute the instructions.

Make the high-level summary of the situation as detailed as necessary to capture all major objects and actions. It is true that the simplified overview does not have all the details and in practice the description is refined until it is complete. The description is complete when enough detail to describe the robot's role and how it will execute that role within the scenario is supplied. Now that Midamba understands exactly what he wants the robot to do, he should extract the vocabulary that is used in the scenario. For example, in Facility Scenario #1, some of the vocabulary is:

- Scan

- Facility

- Inventory

- Return

- Such

- Containers

- Retrieve, and so on

Remember from Chapter 2, "Robot Vocabularies," these types of words represent the scenario and situation vocabulary at the human level and must ultimately be translated into vocabulary that the robot can process. Identifying and removing ambiguity is one of the benefits that can be realized when spelling out the complete scenario.

What does this phrase mean?

If such containers are found retrieve them.

Can it be a little clearer?

Scan the facility.

Is this clear enough to begin the process of converting it to robot language, or is more detail needed? These are subjective questions and will be different depending on experience with converting between human language and robot language.

Breaking Down the Scenario into Situations

In some cases it's easier to first break down a scenario into its sequence of situations and then figure out the particulars for each situation. For example, in Facility Scenario #1, there are several initial situations:

- Robots must report their location within the facility.

- One robot must scan the facility to see what chemicals are available and then report what is found.

- One robot must determine whether one of the chemicals is a match for the job.

- If useful chemicals are found, one of the robots must retrieve the chemicals.

- Code must be uploaded to the robot to change its current programming.

Once you have the scenario broken down into situational components, refine the situation and then identify an appropriate set of commands, variables, actions (i.e., vocabulary) for each situation. Let's look at the situation refinement for the Facility Scenario. Table 11.2 shows the initial description and the first cut at refinement.

Table 11.2   Initial Situation Refinement

Situation Initial Description	Situation Description Refinement
Robots must report their location within the facility.	Using a robot report procedure: Have each robot report its current location within the facility in terms of north, south, east, west of the central office coordinates using centimeters as the unit of distance.
One robot must scan the facility to see what chemicals are available and then report what is found.	Using a robot that is mobile and that can navigate the entire facility, have the robot travel to each area in the facility and have that robot systematically take a series of photos of each area in the facility that has containers of chemicals. Once the robot has traveled to each area and has taken the necessary photos, have the robot use a reporting procedure to make the photos available.
One robot must determine whether one of the chemicals is a match for the job.	Using the photos, determine which chemicals could potentially be useful and then instruct one of the robots to analyze the contents of each candidate chemical. Report the container location and features for any chemicals that meet the search criteria.
If useful chemicals are found, one of the robots must retrieve the chemicals.	If chemicals meeting the search criteria are found, have a robot obtain the locations of each chemical and then travel to each location. One by one retrieve each chemical at that location and return it to the specified coordinates in the front office.
Code must be uploaded to the robot to change its current programming.	Once the necessary robots are identified, upload the necessary code (set of instructions) to reprogram the robot.

Step 2. The Robot's Vocabulary and ROLL Model for Facility Scenario #1

Conceptually, if successful in instructing the robot to carry out its task in each situation, the robot will be able to execute its role for the scenario in its entirety. Now that there is a first cut at the scenario/situation breakdown, the robot's initial level 5 to level 7 ROLL model should be easier to get. Remember the robot's vocabulary at this level is a compromise between human natural language and the robot's Level 1 and level 2 microcontroller language.

 tip

The situations taken together make up the scenario.

Table 11.3 is an initial (and partial) draft of the robot's level 5 through level 7 vocabulary.

Table 11.3 Draft of the Robot's Level 5 Through Level 7 Vocabulary for Facility Scenario #1

Actions	Things
Report	Facility
Navigate, travel	Location
Scan	Centimeter
Systematically take photos	Coordinates, distance, north, south, east, west
Analyze	Current
Retrieve	Area
Obtain location	Containers
Return	Photo
One-by-one retrieve	Chemicals, candidate chemical
Make photos available	Front office coordinates
	Potentially useful search criteria

Midamba will need a more detailed vocabulary as he progresses, but this is a good start. Identifying a potential robot level 5 to level 7 vocabulary at this point serves many purposes. First, it will help him complete the RSVP process for programming the robot. These terms can be used in the flowcharts, statecharts, and area descriptions.

Second, each vocabulary term will ultimately be represented by a variable, class, method, function, or set of procedures in the robot's STORIES code component. So the robot's initial vocabulary gives Midamba a first look at important aspects of the program. Finally, it will help clarify precisely what the robot is to do by helping remove ambiguity and fuzzy ideas. If the robot's instructions and role are not clear, it cannot be reasonably expected that the robot will execute its role following those instructions.

Step 3. RSVP for Facility Scenario #1

Midamba is in a front office or observational room for what appears to be a warehouse. In the process of ransacking the office looking for details on the robots, Midamba came across several floorplan layouts for the facility. He noticed that there were sections marked chemicals on the northwest corner of the facility as well as a section of chemicals located in the southeast corner of the facility. A practical first step of any RSVP is to obtain or generate a visual layout of the area(s) where the robot is to perform its tasks.

Figure 11.1 is a visual layout for the Facility Scenario #1 where the robots are located.

 tip

All the RSVP components and capability matrices for this book were generated with LIBRE office. The area, robot POVs, flowcharts, and statecharts were generated with LIBRE draw. The scenario and situation descriptions were generated using LIBRE writer. We use LIBRE spreadsheet to lay out the capability matrix for each robot.

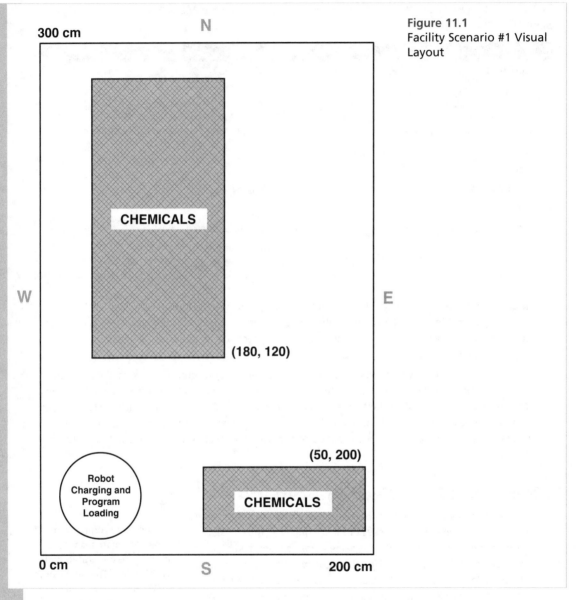

Figure 11.1
Facility Scenario #1 Visual
Layout

🔍 **note**

Each component of the RSVP plays an important role in programming a robot to execute a new task or set of tasks. Once a visual layout of the area and objects the robot is to interact with is developed, one of the most critical steps in the RSVP process is to convert the basic visual layout of the area into what we call a robot point of view (POV) diagram.

Visual Layouts of a Robot POV Diagram

To see what we mean, let's consider the robot's sensors. If the robot has only an ultrasonic sensor and perhaps a color sensor, the robot can only sense things by distance and waves of light. This means the robot can only interact with objects based on their distance or color.

Yes, the robot's interaction with its environment is limited to its sensors, end-effectors, and actuators. One of the primary purposes of generating a layout of the area and objects is to visualize it from your perspective so that you can later represent it in the robot's perspective.

So the robot's POV diagram of the visual layout represents everything the robot interacts with from the perspective of the robot's sensors and capabilities. If all the robot has is a magnetic field sensor, the diagram has to represent everything as some aspect of a magnetic field. If the robot can only take steps in increments of 10 cm, then distances that the robot has to travel within the area have to be represented as some number relative to 10 cm.

If a robot has to retrieve objects made of different materials and different sizes and the robot's end-effectors only have weight, width, pressure, and resistance parameters, all the objects that the robot will interact with in the area have to be described in terms of weight, width, pressure, and resistance. The generation of the visual layout is basically a two-step process:

1. Generate a visual layout from the human perspective.

2. Convert or mark that layout with the robot's POV for everything in the visual layout.

These steps are shown in Figure 11.2.

Figure 11.2
Layout conversion from human to robot's POV

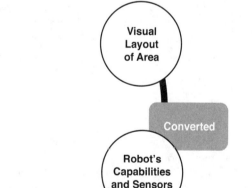

CONVERSION FROM HUMAN TO ROBOT'S POV

Visual
Layout
of Area

Converted

Robot's
Capabilities
and Sensors

Robot POV
Diagram

Once a robot POV diagram has been generated and the robot's action flowchart has been constructed, specifying the robot's SPACES is easier. Figure 11.3 takes the initial visual layout and marks the areas recognizable from the robot's POV.

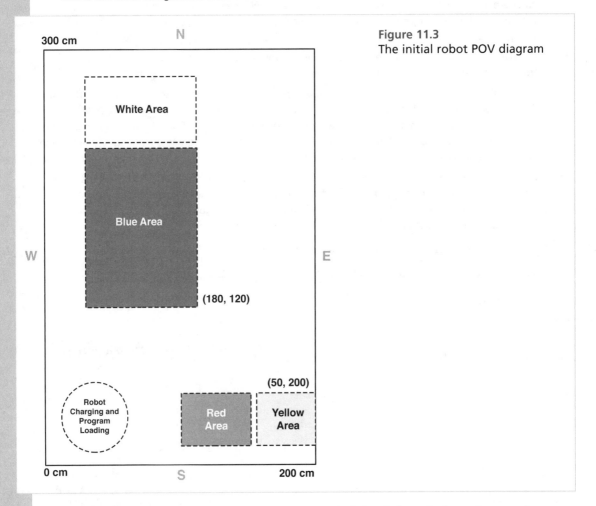

Figure 11.3
The initial robot POV diagram

Notice in Figure 11.3 that there are no longer areas marked chemicals on the layout because the robot currently has no programming designed to identify chemicals. The areas are marked with dotted lines indicating the robot's POV. These areas are marked because that's the only information on the original floorplan that can potentially be recognized by the robot's sensors at this point.

Midamba's Facility Scenario #1 (Refined)

So one of the initial useful situations that can be implemented as a program is for: "Using a robot that is mobile and that can navigate the entire facility, have the robot travel to each area in the facility and systematically take a series of photos of each area in the facility that has containers of chemicals. Once the robot has traveled to each area and has taken the necessary photos, the robot should use a reporting procedure to make the photos available."

Graphical Flowchart Component of the RSVP

If Midamba implements this situation, he will have more information about the warehouse areas. Since Unit2 has a camera attached, Unit2 will be put to work. There is a rough layout of the area (refer to Figure 11.3); a sequence of instructions that Unit2 will execute to accomplish the task as needed. Figure 11.4 is an excerpt from the flowchart (the second component of the RSVP) of the actions that Unit2 should perform.

Figure 11.4
A flowchart excerpt of Unit2's tasks

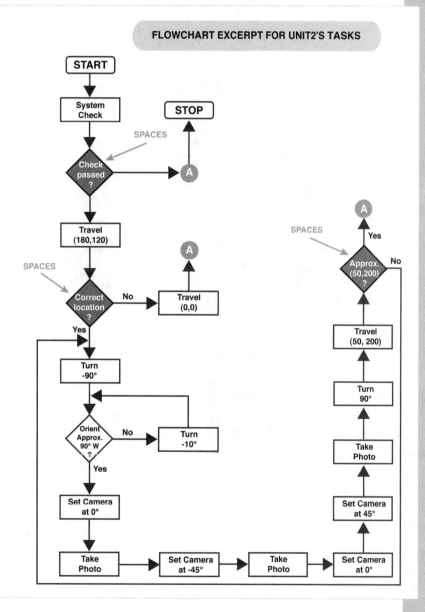

Figure 11.4 is a simplification of the complete flowchart constructed for Unit2. However, it does show the main processing. It shows how Midamba should approach some of the instructions the robot has to execute. From the capability matrix, Unit2 is a bipedal robot equipped with infrared sensors, touch sensors, and a camera. It is driven by a 200 MHz ARM9 and a custom 16-bit processor. The processors have 64 MB and 32 MB, respectively. Figure 11.5 is a photo of Unit2.

PHOTO OF UNIT2

Figure 11.5
Photo of Unit2 (RS Media)

RS MEDIA ROBOT
Bipedal robot
Infrared sensor
Touch sensors
Camera
200 MHz ARM9 Microcontroller
16-bit Processor

The system check is implemented in the constructor. Notice in Figure 11.4 that the SPACES requirement will shut Unit2 down if it does not pass the constructor. As Figure 11.4 shows, the main task of Unit2 is to take photos of what's actually in the warehouse area. Unit2 travels to the specified location and then turns to face the area where chemicals are supposed to be stored and takes a photo. Unit2 is running Linux and uses a Java library for the programming. Although a Java library is used in this instance to program the RS Media, there is a C development tool change available for RS Media that can be downloaded from rsmediadevkit.sourceforge.net.

This is the Java command used to instruct Unit2 to take a photo:

```
System.out.println(Unit2.CAMERA.takePhoto(100));
```

Unit2 has an component named CAMERA, and that CAMERA has a method named takePhoto(). The 100 specifies how long the robot is to pause before actually taking the snapshot. The photo taken is captured in a standard jpeg format. The System.out.println() method has been connected to the SD card in the root. The photo is then stored on the robot's SD card using the System.out.println() method, and it can be retrieved from there.

Notice in the flowchart in Figure 11.4 that the robot's head is adjusted to take photos at different levels. The task that Unit2 performs is often part of the mapping phase of a robotics project. In some cases, the area that the robot will perform in has already been mapped, or there may be floorplans or blueprints that describe the physical area. In other cases, the robot(s) may perform a preliminary surveillance of the area to provide the programmer with enough information to program the primary tasks. In either case, scenario/situation programming requires a detailed understanding of the environment where the robot will operate.

Since Midamba couldn't physically go out into the warehouse area, he needed some idea of what was out there. Unit2 provided Midamba additional information about the warehouse through the photos that it took and uploaded to the computer in the front office. Now Midamba could clearly see that at the floor level, there were glass containers along the northwest corner and southeast corner of the building. The containers seemed to be partially filled with liquids of some type. The containers in the northwest corner had blue labels and some kind of geometric figure, and the containers in southeast corners had yellow labels and geometric figures.

Luckily for Midamba, the containers had no lids and would provide easy access for Unit1's chemical sensors. Midamba also noticed in the northwest corner there appeared to be some kind of electronics on shelves right above the chemicals. If his luck held out, one of the components might be some kind of battery charger. So now that Midamba had a more complete picture of the area, all he needed to do was plan a sequence of instructions for Unit1. This would involve investigating and analyzing the chemicals and the electronic components and retrieving anything that proved to be useful. Figure 11.6 shows a refined robot POV diagram that includes information Unit2 retrieved through photos.

Figure 11.6 shows an area that the robot will be able to navigate and interact with based on distance, color, container sizes, container contents, compass, locations, and level. Notice that the containers, one weighing 119 grams and the other 34 grams, are within the weight capability of Arm1. The diameter of each container, 10 cm and 6 cm, is within the range of Arm1's end-effector grip.

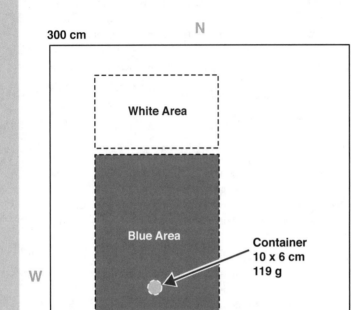

Figure 11.6
A refined POV diagram of the warehouse

Unit1's Tools for the Job

Based on the equipment listed in the capability matrix in Table 11.1, Unit1 can use:

- EV3 ultrasonic sensor to measure distance
- Modified Pixy CMU-5 Arduino camera for object recognition based on color, location, and shape
- Vernier pH sensor to analyze the liquids
- Vernier magnetic sensor in an attempt to find battery charger
- PhantomX Pincher robotic arm (Arm2) to manipulate the sensors
- Tetrix robotic arm (Arm1) to retrieve any useful chemicals or electronic components

Figure 11.7 is the flowchart of the instructions that make up the task `Unit1` is to execute.

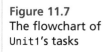

Figure 11.7
The flowchart of
`Unit1`'s tasks

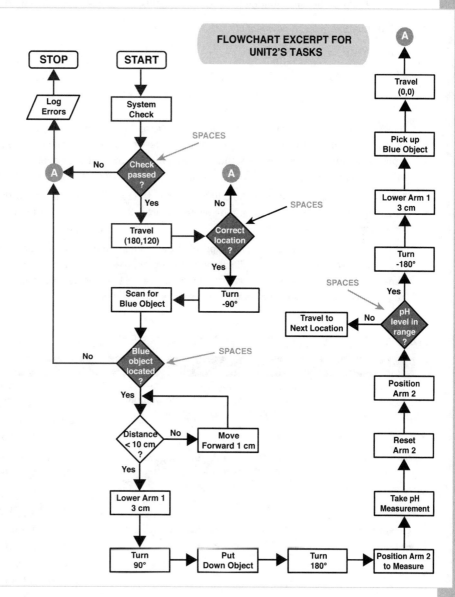

 note

Figure 11.7 is a simplification of `Unit1`'s set of instructions. The actual diagram has 10 pages with far more detail. The complete flowchart for Figure 11.4 and Figure 11.7 is available (along with the complete designs and source code for all the examples of this book) at www.robotteams.org.

Here, we highlight some of the more basic instructions and decisions that the robot has to make. We want to bring particular attention to several of the SPACES checks contained in Figure 11.7:

- Did the robot pass the system check?

- At the correct location?

- pH within range?

- Blue object located?

These types of precondition and postcondition (SPACES) checks are at the heart of programming robots to execute tasks autonomously. We show a few of these decision points in Figure 11.7 for exposition purposes. There were actually dozens for this application. Depending on the application, an autonomous robot program might have hundreds of precondition/postcondition checks.

These decision points represent the critical areas in the robot's programming. If they are not met, it is simply not possible for the robot to execute its tasks. So an important part of the RSVP flowchart component is to highlight the SPACES checks and critical decision points.

Robot "Holy Wars," conflicting schools of robot design and programming, robotics engineering careers made and destroyed are all centered around approaches to handling these kinds of decisions made within an autonomous robot's programming. Table 11.4 generalizes some of the most challenging areas and some of the common approaches to those areas.

Table 11.4   Challenging Areas of Autonomous Robot Programming

Challenging Areas	Description	Approach
Localization	Robot being able to determine where it is within the environment or scenario at any given time	Behavior-based SPA (Sense, Plan, Act) Hierarchical Swarm intelligence
Navigation	Robot having a method of accurately moving from location to location	Dead reckoning Probabilistic localization Coordinate system programming Machine learning
Object recognition	Robot vision; correctly recognizing things in the environment, landmarks, etc.	Image processing Machine learning Neural nets RFID markers
Robot pressure, force	Determining how much pressure or force to apply when lifting, moving, grabbing, placing, or manipulating things in the environment	Machine learning Physics-based programming

Challenging Areas	Description	Approach
Robot control	Determining how the robot will make decisions about what to do next; making decisions about correct localization, navigation, object recognition, end-effector pressure, etc.	Hierarchical Plan-based SPA (Sense, Plan, Act) Behavior-based Bio-inspired Neural networks Swarm intelligence Hybrid

The approaches in Table 11.4 can generally be divided into deliberative and reactive. When we use the term *deliberative* in this book, we mean programming explicitly written by hand (nonautomatic means). We use the term *reactive* to mean instructions learned through machine learning techniques, various puppet-mode approaches, and bio-inspired programming techniques. And of course, there are hybrids of these two basic approaches. The

 tip

There is no one-size-fits-all rule for handling pre/postconditions and assertions.

areas listed in Table 11.4 are focus points of robot programming and where handling the Sensor Precondition Postcondition Assertion Check of Environmental Situations (SPACES) will determine whether the robot can accomplish its tasks.

Everything is dependent on the robot build, the scenario/situation that the robot is in, and the role that the robot will play in the scenario. However, we offer two useful techniques here:

1. At every decision point, use multiple sensors (if possible) for situation verification.

2. Check those sensors against the facts that the robot has established and stored for the situation.

The first technique should involve different types of sensors if possible. For instance, if the robot is supposed to acquire a blue object, you might use a color sensor or camera that identifies the fact that the object is blue. A touch or pressure sensor can be used to verify the fact that the robot actually has grabbed the object, and a compass can be used to determine the robot's position.

The second technique involves using the facts the robot has established about the environment. These facts are also called its *knowledgebase*. If the robot's facts say that the object is supposed to be located at a certain GPS location, is blue, and weighs 34 grams, these facts should be compared against the collective sensor measurements

Collectively, we call these two techniques PASS (propositions and sensor states). The *propositions* are statements or facts that the robot has established to be true about the environment (usually from its original programming or ontology), and the *sensor states* are measurements that have been taken by the robot's sensors and end-effectors. Applying PASS to a situation does not necessarily guarantee that the robot is correct in its localization, navigation, object recognition, and so on, but it does add one more level of confidence to the robot's autonomy.

When looking at the decision points and SPACES in Figure 11.7, keep in mind that in practice it is helpful to consult as many relevant sensors as practical, as well as the robot's knowledgebase, at every critical decision point. In using our approach to storing the information about the scenario in the robot, the facts are stored in the STORIES component introduced in Chapter 10, "An Autonomous Robot Needs STORIES."

State Diagram Component of the RSVP

The state diagram for the RSVP is made easier by the scenario/situation breakdown. Figure 11.8 is the scenario state diagram for Facility Scenario #1.

SCENARIO STATE DIAGRAM OF FACILITY SCENARIO #1

Figure 11.8
The Facility Scenario #1 state diagram.

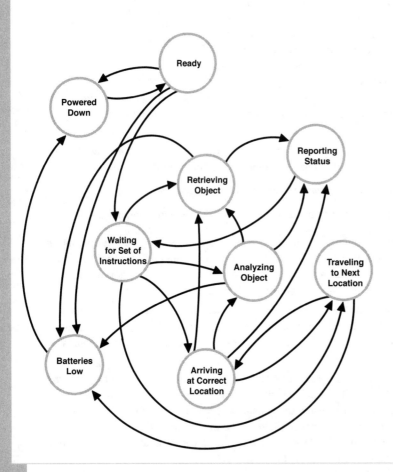

Each circle in Figure 11.8 represents a major situation in the scenario. The robot cannot effectively proceed to the next situation until the SPACES in the current situation are successfully met. We recommend using the PASS technique at each of the SPACES. One situation ordinarily leads to the next, and the situations are usually sequentially dependent, so the robot does not have the option to complete some and not complete others. The state diagram gives us a clear picture of how the robot's autonomy must progress through the scenario as well as what all the major succeed/fail points will be.

Midamba's STORIES for Robot Unit1 and Unit2

Once Midamba was done with the RSVP for his predicament, all he needed to do was develop STORIES for Unit1 and Unit2, upload them, and hopefully he would be on his way. Recall from Chapter 10 one of the major functions of the STORIES component is to capture a description of the situation, objects in the situation, and actions required in the situation as object-oriented code that can be uploaded to the robot. Here we highlight six of the object types:

- Scenario/situation object

- pH sensor object

- Magnetic field sensor object

- Robotic arm object

- Camera object

- Bluetooth/serial communication object

Midamba had to build each of these objects with either Arduino C/C++ code/class libraries or leJOS Java code/class libraries depending on the sensors used.[1] All these objects are part of the STORIES component code that had to be uploaded to Unit1 and Unit2 for the robots to execute our Facility Scenario. These STORIES objects are an improvement over our extended scenario objects discussed in Chapter 10. One of the first improvements is the addition of the scenario class and object shown in BURT Translation Listing 11.1.

BURT Translation Listing 11.1   The scenario **Class**

BURT Translations Output: Java Implementations

```
//Scenario/Situation: SECTION 4
269      class scenario{
270          public ArrayList<situation> Situations;
271          int SituationNum;
272          public scenario(softbot Bot)
273          {
274              SituationNum = 0;
275              situation  Situation1 = new situation(Bot);
276              Situations.add(Situation1);
```

```
277
278
279              }
280          public situation nextSituation()
281          {
282              if(SituationNum < Situations.size()){
283              {
284                  return(Situations.get(SituationNum));
285                  SituationNum++;
286
287              }
288          else{
289
290                      return(null);
291              }
292          }
293
294      }
```

In Chapter 10, we introduced a situation class. A situation may have multiple actions, but what if we have multiple situations? Typically, a scenario can be broken down into multiple situations. Recall from Figure 11.8 that our state diagram divided our warehouse scenario into multiple situations. Recall our situation refinements from Table 11.2. Line 270 in Listing 11.1 is as follows:

```
270          public ArrayList<situation> Situations;
```

This shows that our scenario class can have multiple situations. Using this technique, we can upload complex scenarios to the robot consisting of multiple situations. In fact for practical applications, the scenario object is the primary object declared as belonging to the softbot. All other objects are components of the scenario object. The ArrayList shown on line 270 of Listing 11.1 can contain multiple situation objects. The scenario object accesses the next situation by using the nextSituation() method. In this case, we simply increment the index to get the next situation. But this need not be the case. The nextSituation() method can be implemented using whatever selection criteria is necessary for retrieving the object out of the Situations ArrayList list. BURT Translation Listing 11.2 is an example of one of our situation classes.

BURT Translation Listing 11.2   A situation Class

BURT Translations Output: Java Implementations

```
//Scenario/Situation: SECTION 4
228          class situation{
229
230              public room Area;
```

```
231          int ActionNum = 0;
232          public ArrayList<action>  Actions;
233          action RobotAction;
234          public situation(softbot  Bot)
235          {
236              RobotAction = new action();
237              Actions = new ArrayList<action>();
238              scenario_action1 Task1 = new scenario_action1(Bot);
239              scenario_action2 Task2 = new scenario_action2(Bot);
240              scenario_action3 Task3 = new scenario_action3(Bot);
241              scenario_action4 Task4 = new scenario_action4(Bot);
242              Actions.add(Task1);
243              Actions.add(Task2);
244              Actions.add(Task3);
245              Actions.add(Task4);
246              Area = new room();
247
248          }
249          public void nextAction() throws Exception
250          {
251
252              if(ActionNum < Actions.size())
253              {
254                  RobotAction = Actions.get(ActionNum);
255              }
256              RobotAction.task();
257              ActionNum++;
258
259
260          }
261          public int numTasks()
262          {
263              return(Actions.size());
264
265          }
266
267      }
```

Notice that this situation object consists of an area and a list of actions. Where do we get details for our situations? The RSVP components and the ROLL model Levels 3 through 7 are the sources for the objects that make up each situation. Notice that the situation has an Area declared on line 230 and initialized on line 246. But what does Area consist of? Recall our refined robot POV diagram from Figure 11.6. This diagram gives us the basic components of this situation. BURT Translation Listing 11.3 shows the definition of the room class.

BURT Translation Listing 11.3   Definition of the room Class

BURT Translations Output: Java Implementations

```
//Scenario/Situation: SECTION 4
174     class room{
175         protected int Length = 300;
176         protected int Width = 200;
177         protected int Area;
178         public something BlueContainer;
179         public something YellowContainer;
180         public something  Electronics;
181
182         public  room()
183         {
184             BlueContainer =  new something();
185             BlueContainer.setLocation(180,125);
186             YellowContainer = new something();
187             YellowContainer.setLocation(45,195);
188             Electronics = new something(25,100);
189         }
190
191
192
193         public int   area()
194         {
195             Area = Length * Width;
196             return(Area);
197         }
198
199         public  int length()
200         {
201
202             return(Length);
203         }
204
205         public int width()
206         {
207
208             return(Width);
209         }
210     }
```

Here we show only some of the components of the room for exposition purposes. The room has a size; it contains a blue container and a yellow container and some electronics. Each thing has a location. The containers are declared using the something class from Listing 10.5 in Chapter 10. The scenario, situation, and room classes are major parts of the STORIES component because they are used to describe the physical environment where the robot executes its tasks. They are also used to describe the objects that the robot interacts with. Here, we show enough detail for the reader to understand how these classes must be constructed. Keep in mind that many more details would be needed to fill in the scenario, situation, and area classes. For example, the something class from BURT Translation Listing 10.5 has far more detail. Consider the following:

```
class something{;
    x_location Location;
    int Color;
    float  Weight;
    substance  Material;
    dimensions  Size;
}
```

These attributes, the getter, setter methods as well as the basic error checking, make up only the basics of this class. The more autonomous the robot is, the more detail the scenario, situation, and something classes require. Here the scenarios and situations are kept simple so that the beginner can see and understand the basic structures and coding techniques being used. Every scenario and situation has one or more actions in addition to the things within the scenario. Notice lines 232 to 245 of Listing 11.2; these lines define the actions of the situation. BURT Translation Listing 11.4 shows the declaration of the action class.

BURT Translation Listing 11.4   Declaration of the action Class

BURT Translations Output: Java Implementations

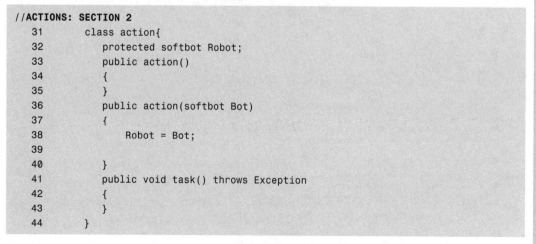

```
//ACTIONS: SECTION 2
31        class action{
32            protected softbot Robot;
33            public action()
34            {
35            }
36            public action(softbot Bot)
37            {
38                Robot = Bot;
39
40            }
41            public void task() throws Exception
42            {
43            }
44        }
```

```
45
46    class scenario_action1  extends action
47    {
48
49       public scenario_action1(softbot Bot)
50       {
51            super(Bot);
52       }
53       public void task() throws Exception
54       {
55            Robot.moveToObject();
56
57       }
58    }
59
60
61
62    class scenario_action2 extends action
63    {
64
65       public  scenario_action2(softbot Bot)
66       {
67
68            super(Bot);
69       }
70
71       public  void task() throws Exception
72       {
73            Robot.scanObject();
74
75       }
76    }
77
78
79
80    class scenario_action3  extends  action
81    {
82
83       public  scenario_action3(softbot Bot)
84       {
85
86            super(Bot);
87       }
88
89       public  void task() throws Exception
90       {
91            Robot.phAnalysisOfObject();
```

```
 92
 93            }
 94        }
 95
 96
 97
 98
 99        class scenario_action4  extends  action
100        {
101
102            public  scenario_action4(softbot Bot)
103            {
104
105                super(Bot);
106            }
107
108            public  void task() throws Exception
109            {
110                Robot.magneticAnalysisOfObject();
111
112            }
113
114
115        }
```

There are four basic actions that we show for this situation. Keep in mind there are more. Action 3 and Action 4 are particularly interesting because they are remotely executed. The code in Listing 11.4 is Java and is running on an EV3 microcontroller, but the phAnalysisofObject() and the magneticAnalysisOfObject() methods are actually implemented by Arduino Uno based components on Unit1. So these methods actually send simple signals through Bluetooth to the Arduino components. Although the implementation language changes, the idea of classes and objects that represent the things in the scenario remains the same. BURT Translation Listing 11.5 shows the Arduino C++ code used to implement the pH analysis, magnetic field analysis, and Bluetooth communication code.

BURT Translation Listing 11.5   Arduino C++ Code for pH and Magnetic Field Analysis and Bluetooth Communication

BURT Translations Output: Arduino C++ Implementations

```
//PARTS: SECTION 1
//Sensor Section
    2    #include <SoftwareSerial.h>    //Software Serial Port
    3    #define RxD 7
    4    #define TxD 6
```

```
5
6    #define DEBUG_ENABLED  1
7
8    SoftwareSerial blueToothSerial(RxD,TxD);
9    // 0.3 setting on Vernier Sensor
10   // measurement unit is Gauss
11
12   class analog_sensor{
13      protected:
14          int Interval;
15          float Intercept;
16          float Slope;
17          float Reading;
18          float Data;
19          float Voltage;
20      public:
21          analog_sensor(float I, float S);
22          float readData(void);
23          float voltage(void);
24          float sensorReading(void);
25
26   };
27   analog_sensor::analog_sensor(float I,float S)
28   {
29       Interval = 3000; // in ms
30       Intercept = I; // in mT
31       Slope = S;
32       Reading = 0;
33       Data = 0;
34   }
35   float analog_sensor::readData(void)
36   {
37       Data = analogRead(A0);
38   }
39   float  analog_sensor::voltage(void)
40   {
41       Voltage = readData() / 1023 * 5.0;
42   }
43
44   float analog_sensor::sensorReading(void)
45   {
46       voltage();
47       Reading = Intercept + Voltage * Slope;
48       delay(Interval);
49       return(Reading);
50   }
51
```

This is the code executed to check the chemicals and electronics that Midamba is searching for. Figure 11.9 shows a photo of Unit1's robot arm holding the pH sensor and analyzing the material in one of the containers. This code is designed to work with Vernier analog sensors, using the Vernier Arduino Interface Shield and an Arduino Uno. In this case, we used a SparkFun RedBoard with an R3 Arduino layout.

Figure 11.9
(a) A photo of Unit1 analyzing a liquid in a container using a pH sensor held by its robotic arm (Arm2).

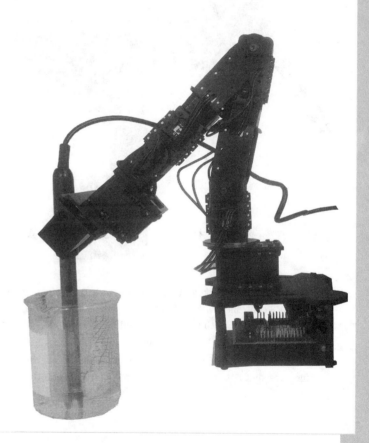

UNIT1's ROBOT ARM (ARM2) HOLDING pH SENSOR

Alkaline battery corrosion is neutralized with substances that have a pH measurement above 7, and nickel battery corrosion is neutralized with substances that have a pH measure below 7. Figure 11.10 shows a photo of Unit1's robot arm holding the magnetic field sensor used to check the electronics for live battery chargers. We chose the 0.3mT (Tesla Setting) using Gauss.

UNIT1's ROBOT ARM (ARM2) HOLDING MAGENTIC FIELD SENSOR

Figure 11.10
(a) A photo of Unit1's robotic arm (Arm2) holding a magnetic field sensor.

The Arduino code takes the measurement and then sends the measurement over Bluetooth back to the EV3 controller where the information is stored as part of the robot's knowledgebase. We used a Bluetooth shield with the Arduino RedBoard and the Vernier shield to accomplish the Bluetooth connection. Figure 11.11 is a photo of the sensor array component that Unit1 uses.

BURT Translation Listing 11.6 shows the main loop of the Arduino controller and how sensor readings are sent both to the serial port and to the Bluetooth connection.

Figure 11.11
The sensor array component connected to the three boards along with the other sensors.

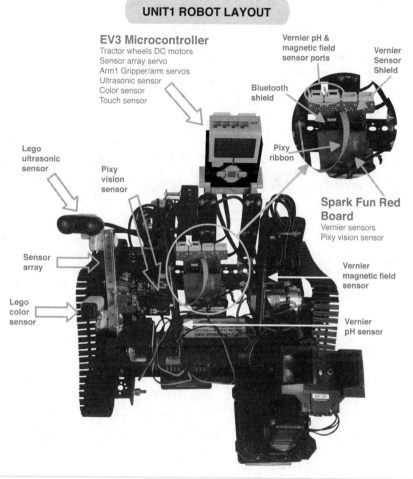

UNIT1 ROBOT LAYOUT

EV3 Microcontroller
Tractor wheels DC motors
Sensor array servo
Arm1 Gripper/arm servos
Ultrasonic sensor
Color sensor
Touch sensor

Vernier pH &
magnetic field
sensor ports

Vernier
Sensor
Shield

Bluetooth
shield

Lego
ultrasonic
sensor

Pixy
vision
sensor

Pixy
ribbon

**Spark Fun Red
Board**
Vernier sensors
Pixy vision sensor

Sensor
array

Vernier
magnetic field
sensor

Lego
color
sensor

Vernier
pH sensor

BURT Translation Listing 11.6   Main Loop of the Arduino Controller

BURT Translations Output: Arduino C++ Implementations

```
//PARTS: SECTION 1
//Sensor Section
53    analog_sensor  MagneticFieldSensor(-3.2,1.6);
54    analog_sensor  PhSensor(13.720,-3.838);
55
56    int ReadingNumber=1;
57
58
59    void setup()
60    {
61        Serial.begin(9600); //initialize serial communication at 9600 baud
```

```
62        pinMode(RxD, INPUT);
63        pinMode(TxD, OUTPUT);
64        setupBlueToothConnection();
65    }
//TASKS: SECTION 3
66    void loop()
67    {
68        float Reading;
69        char InChar;
70        Serial.print(ReadingNumber);
71        Serial.print("\t");
72        Reading = PhSensor.sensorReading();
73        Serial.println(Reading);
74        blueToothSerial.println(Reading);
75        delay(3000);
76        blueToothSerial.flush();
77        if(blueToothSerial.available()){
78            InChar = blueToothSerial.read();
79            Serial.print(InChar);
80        }
81        if(Serial.available()){
82            InChar = Serial.read();
83            blueToothSerial.print(InChar);
84        }
85        ReadingNumber++;
86    }
87    void setupBlueToothConnection()
88    {
89        Serial.println("setting up bluetooth connection");
90        blueToothSerial.begin(9600);
91
92        blueToothSerial.print("AT");
93        delay(400);
94        //Restore all setup values to factory setup
95        blueToothSerial.print("AT+DEFAULT");
96        delay(2000);
97        //set the Bluetooth name as "SeeedBTSlave",the Bluetooth
          //name must be less than 12 characters.
98        blueToothSerial.print("AT+NAMESeeedBTSlave");
99        delay(400);
100       // set the pair code to connect
101       blueToothSerial.print("AT+PIN0000");
102       delay(400);
103
104       blueToothSerial.print("AT+AUTH1");
105       delay(400);
106
107       blueToothSerial.flush();
108   }
```

note

The Bluetooth connection is set up in lines 87 to 107. The transmit pin is set to pin 7 on the Bluetooth shield, and the receive pin is set to pin 6. We used version 2.1 of the Bluetooth shield.

BRON'S

Believe It Or Not!

At the Tenth Anniversary of the Arduino, Programming the Arduino-Compatible Robots Is the Best Way to Go!

The Arduino was released in 2005 with the goal of providing an accessible and easy way for beginners and professionals alike to create microcontroller-based devices that could interact with their environment using sensors and actuators. The Arduino has proven especially useful for building low-cost, entry-level robots in every robotic genre ranging from underwater robots like the OpenRov to the ariel robots such as the ArduoCopter as depicted in Figure 11.12. and everything between.

Figure 11.12

The OpenROV Underwater and the ArduCopter Ariel Open Source Robots based on or compatibile with Arduino

OPEN SOURCE LOW-COST ROBOTS BASED ON/COMPATIBLE WITH ARDUINO

 OPENROV
Telerobotics
Submarine

- **OS:** Linux
- **Hardware:** Open Source
- **CPU:** 720 MHz (BeagleBone ARM Cortex-A8 processor)
- **Memory:** 256 MB DDR2 (BeagleBone)
- **Camera:** HD USB webcam with 2 LED light arrays on servo-tiltable platform
- **Connectivity:** 10 Mb Ethernet data tether
- **Power:** 8 C batteries (~1.5h run time)
- **Dimensions:** 30 cm x 20 cm x 15 cm
- **Weight:** 2.5 kg

ARDUCOPTER
Multirotor /
Helicopter UAV

- **Platform:** Linux, Mac, Windows
- **Hardware:** Open Source
- **Autopilot:** Pixhawk, APM2, Pix4

There are over 1 million Arduino microcontrollers out there, and if you're going to program your robot to interact with the rest of the world, Arduino compatibility is a good place to start. We were able to catch up with Ken Burns, the founder of Tiny Circuits (www.tiny-circuits.com) and the inventor of one of the world's smallest Arduino controllers, as depicted in Figure 11.13.

TINYDUINO COMPONENTS

Figure 11.13
Photo of the Tiny Circuit controller compared to a quarter

The TinyDuino is a miniature open-source electronics platform based on the hardware/software Arduino platform.

Ken describes Tiny Circuits as a company with a focus on "makers and hobbyists that produces a sort-of electronic legos based on the Arduino. That's one of the most used platforms out there, that's the world we play in, and it's also a nice world to work in." In addition to hobbyists and those from various maker spaces, Ken sees the Arduino as a key platform for companies that only require low-volume electronics production. "10 years ago the folks at Arduino were part of the founding fathers of open source hardware." …. "Ten years later the Arduino community is all about making more open source components and making those components easier and more accessible. Things are more cost effective for motor drivers, and the Arduino software base has really matured. Because the Arduino software is so open and accessible, things are easier," Ken stated.

According to Ken "over the next 5 to 10 years personal robotics projects will take off because things are simply getting easier to do." For example, the Arduino is the main platform for many drones and underwater robots. The Underwater Glider is a case in point.

Autonomous Robots to Midamba's Rescue

We highlighted Midamba's RSVP and some of the major components of the robots and softbots he built. The robot, the softbot components, and this approach to programming robots to be autonomous ultimately helped Midamba out of his predicament.

Figure 11.14 shows the story of how Midamba programmed the Unit1 and Unit2 robots to retrieve the neutralizer he needed to clean the corrosion from his battery.

Figure 11.14
How Midamba programmed the Unit1 and Unit2 robots to retrieve the neutralizer.

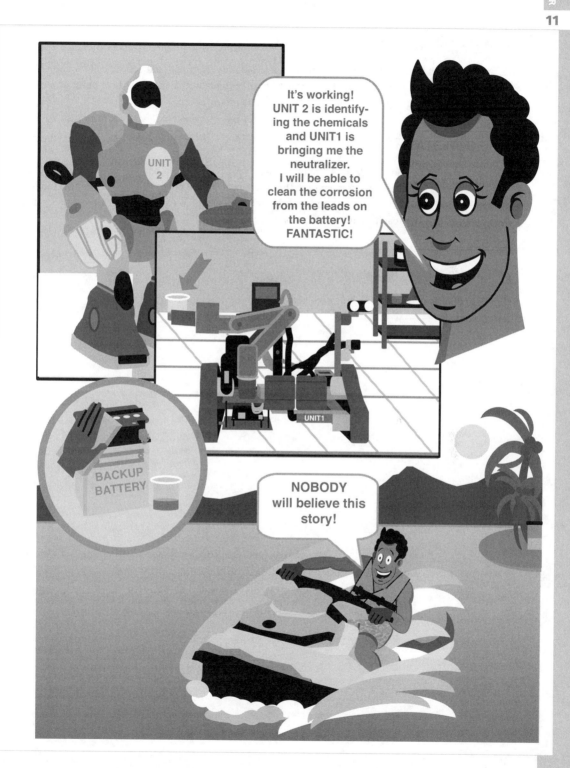

Endnote

1. Although it is possible to program a robot using a single language, our robot projects almost always result in a combination of C++ robot libraries (mostly Arduino) and Java libraries (e.g., Android, leJOS), using sockets and Bluetooth for communication.

What's Ahead?

In Chapter 12, "Open Source SARAA Robots for All!", we wrap up the book by discussing the open source, low-cost robot kits and components that were used in the book. We review the techniques that were used throughout the book and also discuss SARAA (**S**afe **A**utonomous **R**obot **A**pplication **A**rchitecture), our approach to developing autonomous robots.

OPEN SOURCE SARAA ROBOTS FOR ALL!

Robot Sensitivity Training Lesson #12: *Machine learning is no substitute for spending quality programming time with your robot.*

This book provided you with an introductory approach to instructing a robot to execute tasks autonomously using deliberative, scenario-based programming techniques. In addition to being concerned with how to represent the list of instructions to the robot, these approaches focused on

- How to represent the robot's physical environment within the robot's programming

- How to represent the scenario the robot is in within the robot's programming

- How to code the robot's role and actions within the scenario

We introduced you to simple object-oriented and agent-oriented programming techniques as a starting point to address each of the preceding focus areas. We emphasized robot autonomy only within well-understood and predefined scenarios. In particular, we avoided the notion of attempting to program a robot to act autonomously in an environment unknown to the robot and unknown to the programmer.

Introducing you to the concept of programming a robot to execute tasks autonomously has its own set of challenges without complicating matters by adding surprises. Although there are approaches to programming a robot to execute tasks autonomously in an unknown environment, these approaches require advanced robotics knowledge and are beyond the scope of an introductory book.

If you are interested and want to know more about programming robots to handle the unknown, we recommend Ronald Arkin's *Behavior-Based Robotics* and Thomas Braun's *Embedded Robotics: Mobile Robot Design and Applications with Embedded Systems*. If you think you are ready for an intermediate to advanced discussion of our deliberative approach to robot programming we recommend *Agent Technology from a Formal Perspective* by Christopher A. Rouff et al.

Low-Cost, Open-Source, Entry-Level Robots

The robots used in this book were low-cost, entry-level robots. We used the following, as seen in Figure 12.1:

- LEGO EV3 Mindstorms robot controller
- Arduino Uno
- SparkFun Red Board Arduino robot controller
- Trossen's Phantom Pincher robotic arm
- Arduino compatible Arbotix robot controller
- WowWee's RS Media robot with embedded Linux

Figure 12.1
Low cost, open source robot and components used in this book.

LOW COST OPEN SOURCE ROBOTS AND COMPONENTS

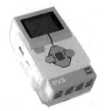

1 EV3 Mindstorms Microcontroller

SOFTWARE
sourceforge.net/p/lejos/wiki/
Developing%20with%20leJOS/

HARDWARE
www.robotshop.com/en/le-go-mindstorms-ev3-us.html

2 Arduino Uno Microcontroller

SOFTWARE
playground.arduino.cc/Main/
SketchList

HARDWARE
store-usa.arduino.cc/

3 Spark Fun Red Board Arduino Microcontroller

SOFTWARE
www.arduino.cc/en/
Tutorial/HomePage

HARDWARE
www.sparkfun.com/
products/12757

4 Phantom X Pincher Robot Arm

HARDWARE
www.trossenrobotics.com/p/Phan-tomX-Pincher-Robot-Arm.aspx

5 ArbotiX Robot Controller

SOFTWARE
code.google.com/p/arbotix/
wiki/BioloidController

6 WowWee RS Media Robot

SOFTWARE
rsmediadevkit.sourceforge.net/

* RS Media no longer available

We also used:

- Arduino Bluetooth shield for communication between controllers

- Pixy (CMUcam5) camera

- Servos and parts from Tetrix

as depicted in Figure 12.2.

Figure 12.2
Low-cost robot components used in this book.

LOW COST ROBOT COMPONENTS

 1 Bluetooth Shield　　 **2** Pixy (CMUcam5) Camera Sensor　　**3** Tetrix Robotics Components

SOFTWARE / HARDWARE
www.seeedstudio.com/depot/
Bluetooth-Shield-p-866.html

SOFTWARE
www.cmucam.org/projects/
cmucam5

HARDWARE
www.tetrixrobotics.com

HARDWARE
charmedlabs.com/default/
pixy-cmucam5/

We used a combination of Vernier, HiTechnic, and LEGO Mindstorms sensors. Our goal was to introduce you to the basics of programming autonomous robots using low-cost, entry-level robots, parts, and sensors. Although we did not use any Raspberry Pi or Beagle Bone-based robot builds, the ideas in this book can be used with any true robot (recall our robot definition from Chapter 1, "What Is a Robot, Anyway?") that has a controller and supports an object-oriented programming language.

Scenario-Based Programming Supports Robot Safety and Programmer Responsibility

We advocate programming robots to act autonomously only within predefined scenarios and situations. If the scenario and situation that the robot is to perform in is well known and understood, safety precautions can be built in from the start. This helps the robot to be safer for interaction with humans, the robot's environment, and other machines. Those of us who program robots have a responsibility to build as many safeguards as necessary to prevent harm to life, the environment, and property. Scenario/situation-based programming helps the programmer to identify and avoid safety pitfalls. While scenario/situation programming is not sufficient alone to prevent safety mishaps, it is a step in the right direction. Regardless of a robot's ultimate set of tasks, if autonomy is involved, safety must be taken into consideration.

SARAA Robots for All

In this book, we introduced you to seven techniques for programming a robot to execute its tasks autonomously:

- Softbot frames

- ROLL models

- REQUIRE

- RSVP

- SPACES

- STORIES

- PASS

Collectively these programming techniques make up what we call SARAA (Safe Autonomous Robot Application Architecture). We call the robots that have this architecture SARAA robots. When implemented correctly, these programming techniques produce a knowledge-based robot controller. Therefore, a SARAA robot is a knowledge-based robot that can act autonomously within preprogrammed scenarios and situations. At Ctest Laboratories (www.ctestlabs.org), SARAA is being designed to work specifically within open source robotics platforms such as Arduino, Linux, and the ROS (Robot Operating System). If the scenarios and situations that SARAA robots are programmed for are well understood and properly defined, then a SARAA robot design helps promote robot safety.

This is true in part because the SPACES and PASS components are specifically designed to address sensor, actuator, end-effector, and robot logic malfunctions, misconstructions, failures, and faults. SARAA robots are context-sensitive by definition. Figure 12.3 shows the basic architecture of a SARAA robot.

Figure 12.3
The basic architecture of a SARAA robot

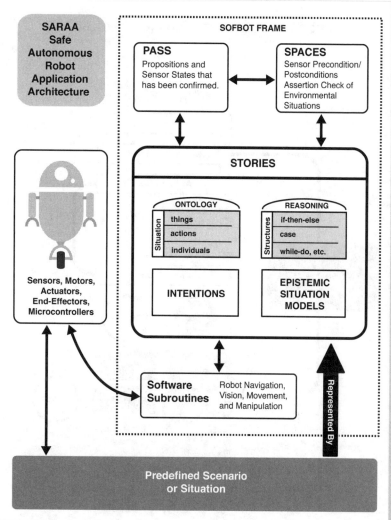

To perform useful tasks, multiple microcontrollers are typically needed for the robot to be fully functional. That is not because we have any specific robot design in mind, but because the nature of things like robot vision, robotic arms, robot navigation, and so on, often require their own dedicated microcontroller. This means that the softbot component shown in Figure 12.3 must have some way to communicate and coordinate the multiple controllers. Figure 12.4 shows the communications architecture for the multiple microcontrollers used.

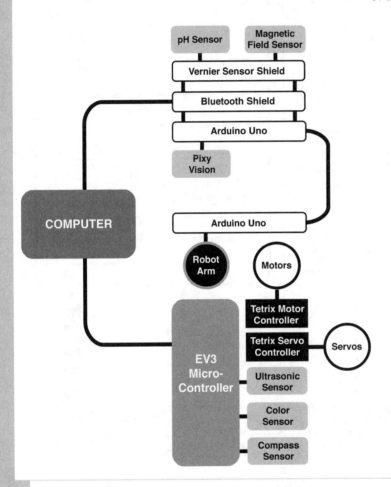

ROBOT'S COMMUNICATION ARCHITECTURE

Figure 12.4
The communications architecture of a SARAA robot

In our robotics lab, we rely primarily on Bluetooth, XBee, and serial communications between the components. All these technologies have open-source implementations. The entire SARAA architecture can be implemented completely within an open source hardware/software environment.

Recommendations for First-Time Robot Programmers

This book was written with a light introduction to SARAA and to programming autonomous robots in general. We kept Midamba's scenario and the other example scenarios/situations simple so that the reader wouldn't get lost in too many details. But to be certain, these were extremely simplified scenarios and situations.

You are encouraged to start with a small project and specific situation/scenario and practice fully implementing (with as much detail as necessary) the RSVP and then the STORIES component for your robot. Start with a single task and then build. Start with a single situation and then add another situation. In this way, you build a library of situations.

Once you have a scenario and all its situations defined and tested, then add another scenario to the robot. Using this approach, you eventually will have a robot that can handle multiple scenarios and many situations. But the key is to start small and build. Be patient. Be thorough.

Complete RSVPs, STORIES, and Source Code for Midamba's Scenario

The complete RSVPs, STORIES, and source code for Midamba's scenario can be downloaded from **www.robotteams.org** along with the actual techniques for neutralizing corrosion on alkaline or nickel-based batteries. In addition to these, we have video of the Unit1 and Unit2 robots autonomously solving Midamba's dilemma.

A

BURT'S GOTCHAS

actuator The motors or devices that power the moving parts of the robot; an actuator could be electric, battery powered, hydraulic, or pneumatic.

agent An entity that can sense or perceive the environment and interact or change the environment in some way. In this book a robot is an agent.

agent-oriented programming The process of composing a list of instructions suitable to be executed by an agent. Robot programming is a form of agent-oriented programming.

android An autonomous robot with a human appearance that has an artificial intelligence–driven controller.

Arduino An open source electronics platform based on easy-to-use hardware and software. It's intended for anyone making interactive projects.

ARM Advanced RISC Machine. One of a family of CPUs based on the RISC (Reduced Instruction Set Computer) architecture.

ARM7 A 32-bit ARM processor.

ARM9 A substantial improvement over the ARM7 processor. It has higher throughput, decreased heat production, and larger cache.

assertion A declarative statement that is purportedly true.

asynchronous Serial data arranged in such a way that timing information is contained within each character rather than being obtained from a master reference. Communication in which timing information is derived from the transmitted information rather than from the terminals or serial lines.

AUAV Autonomous unmanned aerial vehicle.

autonomous In this book, a robot is autonomous when it is not under remote control and its behavior is a consequence of its programming and sensory input.

AUV   Autonomous underwater vehicle.

biped   A robot that has only two legs.

blocking   Waiting for some event to complete before proceeding.

Bluetooth   A global wireless communication standard that connects devices over a certain distance. A Bluetooth connection uses radio waves instead of wires or cables to connect to a device.

BRON   Bluetooth Robot-Oriented Network.

A small team of connected robots that communicate through Bluetooth wireless protocols and the Internet.

BURT   Basic Universal Robotic Translator. Presents code snippets, commands, and robot programs in this book in plain English first.

capability matrix   A table, chart, or spreadsheet that lists the capabilities of a robot in column, row format.

color sensor   A sensor that can measure different electromagnetic wavelengths.

DARPA   Defense Advanced Research Projects Agency.

dead reckoning   In this book it is a method of estimating the position of a robot based on its previous position and its course and speed over a known interval of time.

DOF   Degrees of Freedom. The number of different ways in which the joints of a robot are free to move.

EEPROM   Electronic erasable programmable read-only memory.

end-effector   The device on the end of a robot arm that interacts with or manipulates objects in the robot's environment. The end-effector could take the form of a hand, but it could also take other forms such as drills, wrenches, lasers, clamps, and so on.

episode   In this book, an episode describes a set of stereotyped sequence of events that are part of a commonly known, well-understood narrative (for example, a birthday party has an episode where attendees eat cake, sing songs, give presents, etc.). Episodes, scenarios, and situations are all meant to describe the context where the robot is executing some tasks.

EV3   Microcontroller based on the ARM9 microprocessor from the LEGO Mindstorms line of robotic controllers. It has embedded Linux.

flash RAM   A special type of EEPROM that retains information even when power is turned off.

GNU Linux   A Unix-like computer operating system.

GOTCHAS   Glossary of Technical Concepts and Helpful Acronyms.

gripper   A robot end-effector used for grasping.

heat or temperature sensor   A device that gathers data concerning temperature.

infrared sensor   A light sensor that measures electromagnetic wavelengths longer than 700 nanometers.

interrupt   A break in the normal flow of a system or routine, such that the flow can be resumed from that point at a later time. An interrupt is usually caused by a signal from an external source.

invariant   In programming, an invariant is a condition that can be relied upon to be true during execution of a program, procedure, function, or routine or during some portion of it. It is an assertion that is held to always be true during a certain phase of execution.

leJOS   A tiny Java virtual machine ported to the LEGO NXT microcontroller in 2006. leJOS now has a fully documented robotic API and accompanying set of class libraries.

light sensor   A mechanical or electronic device that measures electromagnetic wavelengths and detects light.

MAC OSX   A Unix-based computer operating system.

microcontroller   A microcomputer on a single integrated circuit used in a control operation or to make changes in a process or operation. All the microcontrollers discussed in this book have sensor capability.

NREF   National Robotics Education Foundation is an informational clearinghouse that identifies the most accessible and affordable curricula, products, and learning resources for robotics education to be used by educators, students, the industry, and the community.

NXT   Microcontroller based on the ARM7 microprocessor from the LEGO Mindstorms line of robotic controllers. It has embedded Linux.

ontology   Definition of types, attributes, characteristics, properties, and relationships of the things and actions for a particular domain of discourse.

OSRF   Open Source Robotics Foundation is an independent non-profit R&D company that supports the development, distribution, and adoption of open source software for use in robotics research, education, and product development.

PASS   Proposition and sensory states. A technique used to verify the execution assumptions of a robot.

pixy (CMUcam5) Vision sensor   A sensor that can be used as part of a vision system for a robot. It was developed as a partnership of Carnegie Mellon and Charmed Labs. Pixy can track objects and connects directly to Arduino and other controllers.

postcondition   A condition, assertion, or proposition about a sequence of logic or value of a variable that must be true immediately after the execution of another piece of code.

precondition   A condition, assertion, or proposition about a sequence of logic or value of a variable that must be true prior to the execution of another piece of code.

quadruped   A robot that has four legs.

READ set   Robot Environmental Attribute Description set. The list of objects that the robot will encounter and interact with in its environment.

reflection   Occurs when light, heat, or sound is bounced off the surface of an object without absorbing it. A reflective sensor can detect and measure the energy that is bounced off the object.

REQUIRE   Robot Effectiveness Quotient Used in Real Environments. Used as an initial litmus test in determining what a robot can and cannot do.

RFID sensor   Radio Frequency IDentification sensor. Used to scan and identify RFID tags, labels, and other devices that store RFID-based information.

robot   A machine that meets the following seven criteria:

1. It must be capable of sensing its external and internal environments in one or more ways through the use of its programming.

2. Its reprogrammable behavior, actions, and control are the result of executing a programmed set of instructions.

3. It must be capable of affecting, interacting with, or operating on its external environment in one or more ways through its programming.

4. It must have its own power source.

5. It must have a language that is suitable for the representation of discrete instructions and data as well as support for programming.

6. Once initiated it must be capable of executing its programming without the need for external intervention (controversial).

7. It must be a nonliving machine.

ROLL model   Robot Ontology Language Level model (see Figure 8.1).

ROV   Remotely Operated Vehicle.

RPA   Remotely Piloted Aircraft.

RS Media   A bipedal biomorphic robot produced by WowWee that has an ARM9 microcontroller and embedded Linux for an operating system; RS Robosapien.

RSVP   Robot Scenario Visual Planning are visuals used to help develop the plan of instructions for what the robot will do. It is composed of a floorplan of the physical environment of the scenario, a statechart of the robot and object's states, and flowcharts of the instructions for the tasks.

SARAA   Safe Autonomous Robot Application Architecture is composed of seven techniques used to program a robot to execute tasks autonomously. These techniques are SOFTBOT Frames, ROLL MODELS, REQUIRE, RSVP, SPACES, STORIES, and PASS.

scenario   A description of a possible course of action within a particular context usually with an expectation of what events should take place, what objects will be encountered, and what the environment will be. Episodes, scenarios, and situations are all meant to describe the context where the robot will be executing some task.

sensor   A device that detects, gathers data, or senses attributes, characteristics, or properties of an internal or external environment.

servo motor   An electric motor combined with an angular position sensor; the length of the control signal sent to the motor determines the angular position of the motor shaft.

situations   All the facts, objects, conditions, and events that affect someone or something within a specified context and environment.

softbot   In this book a software counterpart and softbot representation of the robot implemented as a collection of object-oriented classes.

SPACES   Sensor Precondition/Postcondition Assertion Check of Environmental Situations; used to verify whether it is okay for the robot to carry out its current and next task.

STORIES   Scenarios Translated into Ontologies Reasoning Intentions and Epistemological Situations; the end result of converting a scenario into components that can be represented by object-oriented languages and then uploaded into a robot.

synchronous   A term applied to robots or computers in which the performance of a sequence of operations is controlled by equally spaced clock signals or pulses. Also used to refer to serial data arranged in such a way that timing information is obtained from a master reference rather than for each character.

teleoperation   The operation of a machine by remote control at a distance.

telerobot   A robot controlled by remote control or at a distance.

titanium plated hydraulic powered battle chassis   A frame for an advanced military robot.

torque   The measure of how much a force action on an object causes that object to rotate.

UAV   Unmanned Aerial Vehicle.

ultrasonic sensor   Device that measures distance by using a kind of echolocation using sonar—that is, bouncing sound waves off objects.

INDEX

A

C

G

J-L

S

W-X-Y-Z